John Arthur Thomson

The study of animal life

Second Edition

John Arthur Thomson

The study of animal life
Second Edition

ISBN/EAN: 9783337228675

Printed in Europe, USA, Canada, Australia, Japan

Cover: Foto ©berggeist007 / pixelio.de

More available books at **www.hansebooks.com**

The Study
of
Animal Life

BY

J. ARTHUR THOMSON, M.A., F.R.S.E.

LECTURER ON ZOOLOGY, SCHOOL OF MEDICINE, EDINBURGH
JOINT-AUTHOR OF 'THE EVOLUTION OF SEX'
AUTHOR OF 'OUTLINES OF ZOOLOGY'

SECOND EDITION

WITH ILLUSTRATIONS

LONDON
JOHN MURRAY, ALBEMARLE STREET
1892

"But, for my part, which write the English story, I acknowledge that no man must looke for that at my hands, which I have not received from some other: for I would bee unwilling to write anything untrue, or uncertaine out of mine own invention; and truth on every part is so deare unto me, that I will not lie to bring any man in love and admiration with God and his works, for God needeth not the lies of men.'

TOPSELL'S *Apologia* (1607).

PREFACE

THIS book is intended to help those who would study animal life. From different points of view I have made a series of sketches. I hope that when these are united in the mind of the reader, the picture will have some truth and beauty.

My chief desire has been to give the student some impulse to joyousness of observation and freedom of judgment, rather than to satisfy that thirst for knowledge which leads many to intellectual insobriety. In pursuance of one of the aims of this series, I have also tried to show how our knowledge of animal life has grown, and how much room there is for it still to grow.

A glance at the table of contents will show the plan of the book; first, the everyday life of animals, next, their internal activities, thirdly, their forms and structure, and finally, the theory of animal life. This is a commonly accepted mode of treatment, and it is one by which it is possible in different parts of the book to appeal to students of different tastes. For, in lecturing to those who attend University Extension Courses, I find that seniors are most interested in the general problems of evolution, heredity, and environment; that others care more about the actual forms of life and their structure; that many desire to have a clear understanding of the functions of the animal body; while most wish to study the ways of living animals, their struggles and loves, their homes and societies. To each of these classes of students a quarter of this volume is dedicated; perhaps they will correct their partiality by reading the whole.

As to the two Appendixes, I may explain that instead of giving references at the end of each chapter, I have combined these in a connected bibliography; the other Appendix on "Animal Life and Ours" may show how my subject is related to some of the others usually discussed in University Extension Courses.

My friend Mr. Norman Wyld has written the three chapters on "The Powers of Life," pp. 125-166, and I am also indebted to him for helpful suggestions in regard to other parts of the book. I have to thank Mr. Murray, Messrs. Chambers, and Mr. Walter Scott, for many of the illustrations; while several original drawings have been made for me by my friend Mr. William Smith.

Professor Knight and Mr. John Murray have given me many useful hints while the book was passing through the press, and Mr. Ricardo Stephens was good enough to read the proof sheets.

J. A. T.

SCHOOL OF MEDICINE,
    EDINBURGH, *May* 1892

CONTENTS

PART I

THE EVERYDAY LIFE OF ANIMALS

CHAPTER I

THE WEALTH OF LIFE

1. *Variety of life*—2. *Haunts of life*—3. *Wealth of form*—4. *Wealth of numbers*—5. *Wealth of beauty* . . . . Pages 1-17

CHAPTER II

THE WEB OF LIFE

1. *Dependence upon surroundings*—2. *Inter-relations of plants and animals*—3. *Relation of animals to the earth*—4. *Nutritive relations*—5. *More complex interactions* . . . 18-31

CHAPTER III

THE STRUGGLE OF LIFE

1. *Nature and extent of the struggle*—2. *Armour and weapons*—3. *Different forms of struggle*—4. *Cruelty of the struggle* . 32-45

CHAPTER IV

SHIFTS FOR A LIVING

1. *Insulation*—2. *Concealment*—3. **Parasitism**—4. *General resemblance to surroundings*—5. **Variable colouring**—6. *Rapid change of colour*—7. *Special protective resemblance*—8. *Warning colours*—9. *Mimicry*—10. *Masking*—11. *Combination of advantageous qualities*—12. *Surrender of parts* . . Pages 46-66

CHAPTER V

SOCIAL LIFE OF ANIMALS

1. *Partnerships*—2. *Co-operation and division of labour*—3. *Gregarious life and combined action*—4. *Beavers*—5. *Bees*—6. *Ants*—7. *Termites*—8. *Evolution of social life*—9. *Advantages of social life*—10. *A note on "the social organism"*—11. *Conclusions* 67-94

CHAPTER VI

THE DOMESTIC LIFE OF ANIMALS

1. **The love of mates**—2. *Love and* **care for offspring** . 95-116

CHAPTER VII

THE INDUSTRIES OF ANIMALS

1. *Hunting*—2. *Shepherding*—3. *Storing*—4. *Making of homes*—5. *Movements* 117-124

PART II

THE POWERS OF LIFE

CHAPTER VIII

VITALITY

1. *The task of physiology*—2. *The seat of life*—3. *The energy of life*—4. *Cells, the elements of life*—5. *The machinery of life*—6. *Protoplasm*—7. *The chemical elements of life*—8. *Growth*—9. *Origin of life* . . . . Pages 125-142

CHAPTER IX

THE DIVIDED LABOURS OF THE BODY

1. *Division of labour*—2. *The functions of the body: Movement; Nutrition; Digestion; Absorption; The work of the liver and the kidneys; Respiration; Circulation; The changes within the cells; The activities of the nervous system*—3. *Sketch of Psychology* 143-152

CHAPTER X

INSTINCT

1. *General usage of the term*—2. *Careful usage of the term*—3. *Examples of instinct*—4. *The origin of instinct* . 153-166

PART III

THE FORMS OF ANIMAL LIFE

CHAPTER XI

THE ELEMENTS OF STRUCTURE

1. *The resemblances and contrasts between plants and animals*—2. *The relation of the simplest animals to those which are more complex*—3. *The parts of the animal body* . . Pages 167-183

CHAPTER XII

THE LIFE-HISTORY OF ANIMALS

1. *Modes of reproduction*—2. *Divergent modes*—3. *Historical*—4. *The egg-cell or ovum*—5. *The male-cell or spermatozoon*—6. *Maturation of the ovum*—7. *Fertilisation*—8. *Segmentation and the first stages in development*—9. *Some generalisations: the ovum theory, the Gastræa theory, fact of recapitulation, organic continuity* 184-203

CHAPTER XIII

THE PAST HISTORY OF ANIMALS

1. *The two records*—2. *Imperfection of the geological record*—3. *Palæontological series*—4. *Extinction of types*—5. *Various difficulties*—6. *Relative antiquity of animals* . . 204-209

CHAPTER XIV

THE SIMPLEST ANIMALS

1. *The simplest forms of life*—2. *Survey of Protozoa*—3. *The common Amœba*—4. *Structure of the Protozoa*—5. *Life of Protozoa*—6. *Psychical life of the Protozoa*—7. *History of the Protozoa*—8. *Relation to the earth*—9. *Relation to other forms of life*—10. *Relation to man* 210-221

CHAPTER XV

BACKBONELESS ANIMALS

1. *Sponges*—2. *Stinging-animals or Cœlenterata*—3. *"Worms"*—4. *Echinoderms*—5. *Arthropods*—6. *Molluscs* . Pages 222-247

CHAPTER XVI

BACKBONED ANIMALS

1. *Balanoglossus*—2. *Tunicates*—3. *The Lancelet*—4. *Round-mouths or Cyclostomata* — 5. *Fishes* — 6. *Amphibians* — 7. *Reptiles* — 8. *Birds*—9. *Mammals* 248-272

PART IV

THE EVOLUTION OF ANIMAL LIFE

CHAPTER XVII

THE EVIDENCES OF EVOLUTION

1. *The idea of evolution*—2. *Arguments for evolution: Physiological, Morphological, Historical*—3. *Origin of life* . . 273-281

CHAPTER XVIII

THE EVOLUTION OF EVOLUTION THEORIES

1. *Greek philosophers*—2. *Aristotle*—3. *Lucretius*—4. *Evolutionists before Darwin*—5. *Three old masters: Buffon, Erasmus Darwin, Lamarck*—6. *Darwin*—7. *Darwin's fellow-workers*—8. *The present state of opinion* 282-302

CHAPTER XIX

THE INFLUENCE OF HABITS AND SURROUNDINGS

1. *The influence of function*—2. *The influence of surroundings*—3. *Our own environment* 303-319

CHAPTER XX

HEREDITY

1. *The facts of heredity*—2. *Theories of heredity, historical retrospect*—3. *The modern theory of heredity*—4. *The inheritance of acquired characters*—5. *Social and ethical aspects*—6. *Social inheritance*

Pages 320-339

APPENDIX I

ANIMAL LIFE AND OURS

A. Our relation to animals: 1. *Affinities and differences between man and monkeys*—2. *Descent of man*—3. *Various opinions about the descent of man*—4. *Ancestors of man*—5. *Possible factors in the ascent of man.* B. Our relation to Biology: 6. *The utility of science*—7. *Practical justification of biology*—8. *Intellectual justification of biology* 340-350

APPENDIX II

SOME OF THE BEST BOOKS ON ANIMAL LIFE

A. *Books on "Zoology"*—B. *Books on "Natural History"*—C. *Books on "Biology"* 351-369

INDEX 371-375

PART I

THE EVERYDAY LIFE OF ANIMALS

CHAPTER I

THE WEALTH OF LIFE

1. *Variety of Life*—2. *Haunts of Life*—3. *Wealth of Form*—
4. *Wealth of Numbers*—5. *Wealth of Beauty*

THE first steps towards an appreciation of animal life must be taken by the student himself, for no book-lore can take the place of actual observation. The student must wash the quartz and dig for the diamonds, though a book may help him to find these, and thereafter to fashion them into a treasure.

Happily, however, the raw material of observation is not rare like gold or diamonds, but near to us as sunshine and rain-drops. Within a few hours' walk of even the largest of our towns the country is open and the animals are at home. Though we may not be able to see "the buzzard homing herself in the sky, the snake sliding through creepers and logs, the elk taking to the inner passes of the woods, or the razor-billed auk sailing far north to Labrador," we can watch our own delightful birds building their "homes without hands," we can study the frogs from the time that they trumpet in the early spring till they or their offspring seek winter quarters in the mud, we can follow the bees and detect their adroit burglary of the flowers. And if we are discontented with our opportunities, let us read Gilbert White's *History of Selborne*, or how Darwin watched earthworms for half a lifetime, or how Richard Jefferies saw in the fields and hedgerows of Wiltshire a vision of nature, which seemed every year to grow richer in beauty and marvel. It is thus that the study of Natural History should begin, as it does naturally begin in childhood, and as it began long before there was any exact Zoology,—with the observation of animal life in its familiar forms. The country schoolboy, who watches the squirrels hide the beech nuts and pokes the hedgehog into a living ball, who finds the nest of the lapwings, though they decoy him away with prayerful cries, who catches the speckled trout in spite of all their caution, and laughs at the ants as they expend hours of labour on booty not worth the having, is laying the foundation of a naturalist's education, which, though he may never build upon it, is certainly the surest. For it is in such studies that we get close to life, that we may come to know nature as a friend, that we may even hear the solemn beating of her heart.

The same truth has been vividly expressed by one whose own life-work shows that thoroughness as a zoologist is consistent with enthusiasm for open-air natural history. Of the country lad Dr. C. T. Hudson says, in a Presidential Address to the Royal Microscopical Society, that he "wanders among fields and hedges, by moor and river, sea-washed cliff and shore, learning zoology as he learnt his native tongue, not in paradigms and rules, but from Mother Nature's own lips. He knows the birds by their flight and (still rarer accomplishment) by their cries. He has never heard of *Œdicnemus crepitans*, the *Charadrius pluvialis*, or the *Squatarola cinerea*, but he can find a plover's nest, and has seen the young brown peewits peering at him from behind their protecting clods. He has watched the cunning flycatcher leaving her obvious and yet invisible young in a hole in an old wall, while she carries off the pellets that might betray their presence; and has stood so still to see the male redstart that a field-mouse has curled itself on his warm foot and gone to sleep."

But the student must also attempt more careful studies of living animals, for it is easy to remain satisfied with vague "general impressions." He should make for himself —to be corrected afterwards by the labours of others—a "Fauna" and "Flora" of the district, or a "Naturalist's Year Book" of the flow and ebb of the living tide. He should select some nook or pool for special study, seeking a more and more intimate acquaintance with its tenants, watching them first and using the eyes of other students afterwards. Nor is there any difficulty in keeping at least freshwater aquaria—simply glass globes with pond water and weeds—in which, within small compass, much wealth of life may be observed. Those students are specially fortunate who have within reach such collections as the Zoological Gardens and the British Museum in London; but this is no reason for failing to appreciate the life of the sea-shore, the moor-pond, and the woods, or for neglecting to gain the confidence of fishermen and gamekeepers, or of any whose knowledge of natural history has been gathered from the experience of their daily life.

1. **Variety of Life.**—Between one form of life and another there often seems nothing in common save that both are alive. Thus life is characteristically asleep in plants, it is generally more or less awake in animals. Yet among the latter, does it not doze in the tortoise, does it not fever in the hot-blooded bird? Or contrast the phlegmatic amphibian and the lithe fish, the limpet on the rock and the energetic squid, the barnacle passively pendent on the floating log and the frolicsome shrimp, the cochineal insect like a gall upon the leaf and the busy bee, the sedentary corals and the free-swimming jellyfish, the sponge on the rock and the minute Night-Light Infusorians which make the waves sparkle in the summer darkness. No genie of Oriental fancy was more protean than the reality behind the myth —the activity of life.

2. **Haunts of Life.**—The variety of haunt and home is not less striking. There is the great and wide sea with swimming things innumerable, our modern giants the whales, the seals and walruses and the sluggish sea-cows, the flip-

pered penguins and Mother Carey's chickens, the marine turtles and swift poisonous sea-serpents, the true fishes in prolific shoals, the cuttles and other pelagic molluscs; besides hosts of armoured crustaceans, swiftly-gliding worms, fleets of Portuguese Men-of-War and throbbing jellyfish, and minute forms of life as numerous in the waves as motes in the sunlit air of a dusty town.

> "But what an endless worke have I in hand,
> To count the seas abundant progeny,
> Whose fruitful seede farre passeth those on land,
> And also those which wonne in th' azure sky;
> For much more eath to tell the starres on hy,
> Albe they endlesse seem in estimation,
> Then to recount the seas posterity;
> So fertile be the flouds in generation,
> So huge their numbers, and so numberlesse their nation."

Realise Walt Whitman's vivid picture:—

> "The World below the brine.
> Forests at the bottom of the sea—the branches and leaves,
> Sea-lettuce, vast lichens, strange flowers and seeds—the thick tangle, the openings, and the pink turf,
> Different colours, pale grey and green, purple, white, and gold—the play of light through the water,
> Dumb swimmers there among the rocks—coral, gluten, grass, rushes—and the aliment of the swimmers,
> Sluggish existences grazing there, suspended, or slowly crawling close to the bottom:
> The sperm-whale at the surface, blowing air and spray, or disporting with his flukes,
> The leaden-eyed shark, the walrus, the turtle, the hairy sea-leopard, and the sting ray.
> Passions there, wars, pursuits, tribes—sight in those ocean depths—breathing that thick breathing air, as so many do."

The sea appears to have been the cradle, if not the birthplace, of the earliest forms of animal life, and some have never wandered out of hearing of its lullaby. From the sea, animals seem to have migrated to the shore and thence to the land, but also to the great depths. Of the life of the deep sea we have had certain knowledge only

FIG. 1.—Suggestion of deep-sea life. (In part from a figure by W. Marshall.)

within the last quarter of a century, since the *Challenger* expedition (1872-76), under Sir Wyville Thomson's leadership, following the suggestions gained during the laying of the Atlantic cables and the tentative voyages of the *Lightning* (1868) and the *Porcupine* (1870), revealed what was virtually a new world. During $3\frac{1}{2}$ years the *Challenger* explorers cruised over 68,900 nautical miles, reached with the long arm of the dredge to depths equal to reversed Himalayas, raised sunken treasures of life from over 300 stations, and brought home spoils which for about twenty years have kept the savants of Europe at work, the results of which, under Dr. John Murray's editorship, form a library of about forty huge volumes. The discovery of this new world has not only yielded rich treasures of knowledge, but has raised a wave of wider than national enthusiasm which has not since died away.

We are at present mainly interested in the general picture which the results of these deep-sea explorations present,—of a thickly-peopled region far removed from direct observation, sometimes three to five miles beneath the surface—a world of darkness relieved only by the living lamps of phosphorescence, of silent calm in which animals grow into quaint forms of great uniformity throughout wide areas, and moreover a cold and plantless world in which the animals have it all their own way, feeding, though apparently without much struggle for existence, on their numerous neighbours, and ultimately upon the small organisms which in dying sink gently from the surface like snowflakes through the air.

Far otherwise is it on the shore—sunlight and freshening waves, continual changes of time and tide, abundant plants, crowds of animals, and a scrimmage for food. The shore is one of the great battlefields of life on which, through campaign after campaign, animals have sharpened one another's wits. It has been for untold ages a great school.

Leaving the sea-shore, the student might naturally seek to trace a migration of animals from sea to estuary, and from the brackish water to river and lake. But this path, though followed by some animals, does not seem to have been that which led to the establishment of the greater part of our freshwater fauna. Professor Sollas has shown with much conclusiveness that the conversion of comparatively shallow continental seas into freshwater lakes has taken place on a large scale several times in the history of the earth. This has been in all likelihood accompanied by the transformation of marine into freshwater species. It is thus, we believe, that our lakes and rivers were first peopled. Many freshwater forms differ from their marine relatives in having suppressed the obviously hazardous free-swimming juvenile stages, in bearing young which are sedentary or in some way saved from being washed away by river currents. Minute and lowly, but marvellously entrancing, are numerous Rotifers, of which we know much through the labours of Hudson and Gosse. These minute forms are among the most abundant tenants of fresh water, and their eggs are carried from one watershed to another on the wings of the wind and on the feet of birds, so that the same kinds may be found in widely separate waters. Let us see them in the halo of Hudson's eulogy: "To gaze into that wonderful world which lies in a drop of water, crossed by some atoms of green weed; to see transparent living mechanism at work, and to gain some idea of its modes of action; to watch a tiny speck that can sail through the prick of a needle's point; to see its crystal armour flashing with ever-varying tints, its head glorious with the halo of its quivering cilia; to see it gliding through the emerald stems, hunting for its food, snatching at its prey, fleeing from its enemy, chasing its mate (the fiercest of our passions blazing in an invisible speck); to see it whirling in a mad dance to the sound of its own music—the music of its happiness, the exquisite happiness of living,—can any one who has once enjoyed this sight, ever turn from it to mere books and drawings, without the sense that he has left all fairyland behind him?" Not less lively than the Rotifers are crowds of minute crustaceans or water-fleas which row swiftly through the clear water, and are eaten in hundreds by the fishes. But there are higher forms still; crayfish, and the larvæ of mayflies and dragonflies, mussels and water-snails, fishes and newts, the dipper and the kingfisher, the otter and the vole.

As we review the series of animals from the simplest upwards, we find a gradual increase in the number of those which live on land. The lowest animals are mostly aquatic—the sponges and stinging-animals wholly so; worm-like forms which are truly terrestrial are few compared with those in water; the members of the starfish group are wholly marine; among crustaceans, the woodlice, the land-crabs, and a few dwellers on the land, are in a small minority; among centipedes, insects, and spiders the aquatic forms are quite exceptional; and while the great majority of molluscs live in water, the terrestrial snails and slugs are legion. In the series of backboned animals, again, the lowest forms are wholly aquatic; an occasional fish like the climbing-perch is able to live for a time ashore; the mud-fish, which can survive being brought from Africa to Europe within its dry "nest" of mud, has learned to breathe in air as well as in water; the amphibians really mark the transition from water to dry land, and usually rehearse the story in each individual life as they grow from fish-like tadpoles into frog- or newt-like adults. Among reptiles, however, begins that possession of the earth, which in mammals is established and secure. As insects among the backboneless, so birds among the backboned, possess the air, achieving in perfection what flying fish, swooping treefrogs and lizards, and above all the ancient and extinct flying reptiles, have reached towards. Interesting, too, are the exceptions—ostriches and penguins, whales and bats, the various animals which have become burrowers, the dwellers in caves, and the thievish parasites.

But it is enough to emphasise the fact of a general ascent from sea to shore, from shore to dry land, and eventually into the air, and the fact that the haunts and homes of animals are not less varied than the pitch of their life.

3. **Wealth of Form.**—As our observations accumulate, the desire for order asserts itself, and we should at first classify for ourselves, like the savage before us, allowing similar impressions to draw together into groups, such as

birds and beasts, fishes and worms. At first sight the types of architecture seem confusingly numerous, but gradually certain great samenesses are discerned. Thus we distinguish as *higher* animals those which have a supporting rod along the back, and a nerve cord lying above this; while the *lower* animals have no such supporting rod, and have their nerve-cord (when present) on the under, not on the upper side of the body. The higher or backboned series has its double climax in the Birds and the furred Mammals. Indissolubly linked to the Birds are the Reptiles,—lizards and snakes, tortoises and crocodiles—the survivors of a great series of ancient forms, from among which Birds, and perhaps Mammals also, long ago arose. Simpler in many ways, as in bones and brains, are Amphibians and Fishes in close structural alliance, with the strange double-breathing, gill- and lung-possessing mud-fishes as links between them. Far more old-fashioned than Fishes, though popularly included along with them, are the Round-mouths—the half-parasitic hag-fish, and the palatable lampreys, with quaint young sometimes called "nine-eyes." Near the base of this series is the lancelet, a small, almost translucent animal living in the sea-sand at considerable depths. It may be regarded as a far-off prophecy of a fish. Just at the threshold of the higher school of life, the sea-squirts or Tunicates have for the most part stumbled; for though the active young forms have been acknowledged for many years as reputable Vertebrates, almost all the adults fall from this estate, and become so degenerate that no zoologist ignorant of their life-history would recognise their true position. Below this come certain claimants for Vertebrate distinction, notably one *Balanoglossus*, a worm-like animal, idolised by modern zoology as a connecting link between the backboned and backboneless series, and reminding us that exact boundary-lines are very rare in nature. For our present purpose it is immaterial whether this strange animal be a worm-like vertebrate or a vertebrate-like worm.

Across the line, among the backboneless animals, it is more difficult to distinguish successive grades of higher

and lower, for the various classes have progressed in very different directions. We may liken the series to a school in which graded standards have given place to classes which have "specialised" in diverse studies; or to a tree whose branches, though originating at different levels, are all strong and perfect. Of the shelled animals or Molluscs there are three great sub-classes, (*a*) the cuttlefishes and the pearly nautilus, (*b*) the snails and slugs, both terrestrial and aquatic, and (*c*) the bivalves, such as cockle and mussel, oyster and clam. Simpler than all these are a few forms which link molluscs to worms.

Clad in armour of a very different type from the shells of most Molluscs are the jointed-footed animals or Arthropods, including on the one hand the almost exclusively aquatic crustaceans, crabs and lobsters, barnacles and "water-fleas," and on the other hand the almost exclusively aerial or terrestrial spiders and scorpions, insects and centipedes, besides quaint allies like the "king-crab," the last of a strong race. Again a connecting link demands special notice, *Peripatus* by name, a caterpillar- or worm-like Arthropod, breathing with the air-tubes of an insect or centipede, getting rid of its waste-products with the kidneys of a worm. It seems indeed like "a surviving descendant of the literal father of flies," and suggests forcibly that insects rose on wings from an ancestry of worms much as birds did from the reptile stock.

Very different from all these are the starfishes, brittle-stars, feather-stars, sea-urchins, and sea-cucumbers, animals mostly sluggish and calcareous, deserving their title of thorny-skinned or Echinodermata. Here again, moreover, the sea-cucumbers or Holothurians exhibit features which suggest that this class also originated from among "worms."

But "Worms" form a vast heterogeneous "mob," heart-breaking to those who love order. No zoologist ever speaks of them now as a "class"; the title includes many classes, bristly sea-worms and the familiar earthworms, smooth suctorial leeches, ribbon-worms or Nemerteans, round hair-worms or Nematodes, flat tapeworms and flukes, and many others with hardly any characters in common. To us these many kinds of "worms" are full of interest, because in the past they must have been rich in progress, and zoologists find among them the bases of the other great branches—Vertebrates, Molluscs, Arthropods, and Echinoderms. "Worms" lie in a central (and still muddy) pool, from which flow many streams.

Lower still, and in marked contrast to the rest, are the Stinging-animals, such as jellyfish throbbing in the tide, zoophytes clustering like plants on the rocks, sea-anemones like bright flowers, corals half-smothered with lime. In the Sponges the type of architecture is often very hard to find. They form a branch of the tree of life which has many beautiful leaves, but has never risen far.

Beyond this our unaided eyes will hardly lead us, yet the pond-water held between us and the light shows vague specks like living motes, the firstlings of life, the simplest animals or Protozoa, almost all of which have remained mere unit specks of living matter.

It is easy to write this catalogue of the chief forms of life, and yet easier to read it: to have the tree of life as a living picture is an achievement. It is worth while to think and dream over a bird's-eye view of the animal kingdom—to secure representative specimens, to arrange them in a suitably shelved cupboard, so that the outlines of the picture may become clear in the mind. The arrangement of animals on a genealogical or pedigree tree, which Haeckel first suggested, may be readily abused, but it has its value in presenting a vivid image of the organic unity of the animal kingdom.

If the catalogue be thus realised, if the foliage come to represent animals actually known, and if an attempt be made to learn the exact nature, limits, and meaning of the several branches, the student has made one of the most important steps in the study of animal life. Much will remain indeed—to connect the living twigs with those whose leaves fell off ages ago, to understand the continual renewal of the foliage by the birth of new leaves, and finally to understand how the entire tree of life grew to be what it is.

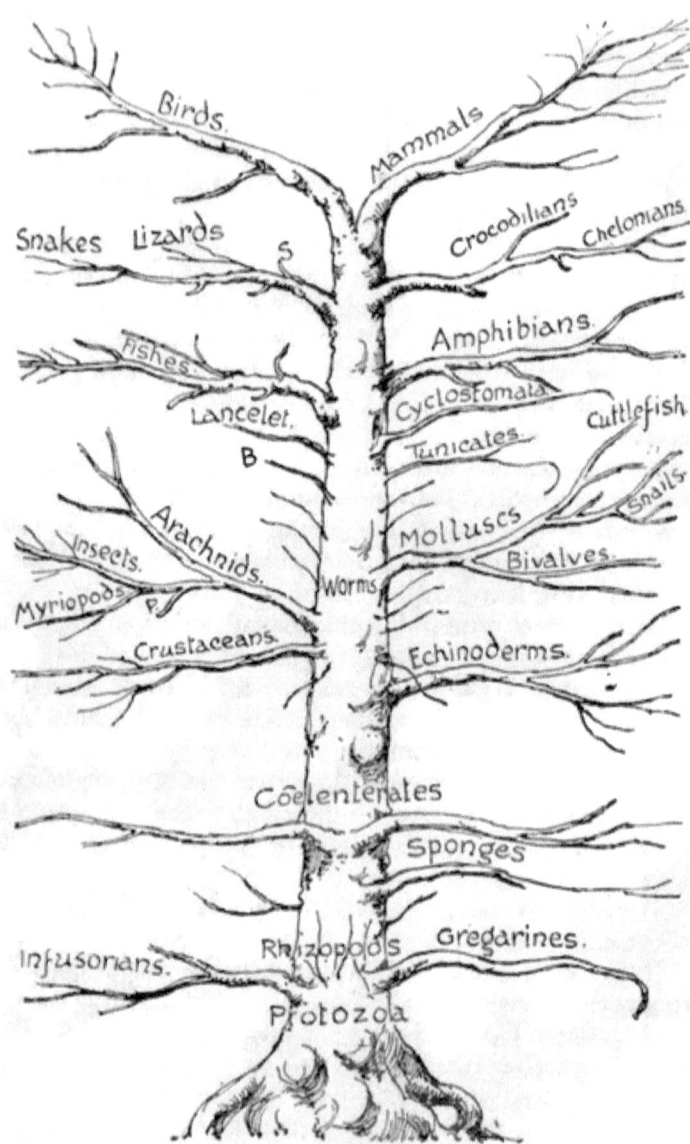

FIG. 2 — Genealogical Tree

The small branches in the centre indicate the classes of "worms"; the letters P, B, and S indicate the positions of Peripatus, Balanoglossus, and Sphenodon or Hatteria respectively.

There is of course no doubt as to the fact that some forms of life are more complex than others. It requires no faith to allow that the firstlings or Protozoa are simpler than all the rest; that sponges, which are more or less loose colonies of unit masses imperfectly compacted together, are in that sense simpler than jellyfish, and so on. The animals most like ourselves are more intricate and more perfectly controlled organisms than those which are obviously more remote, and associated with this perfecting of structure there is an increasing fulness and freedom of life.

We may arrange all the classes in series from low to high, from simple to complex, but this will express only our most generalised conceptions. For within each class there is great variety, each has its own masterpieces. Thus the simplest animals are often cased in shells of flint or lime whose crystalline architecture has great complexity. The simplest sponge is little more than a double-walled sack riddled by pores through which the water is lashed, but the Venus' Flower-Basket (*Euplectella*), one of the flinty sponges, has a complex system of water canals and a skeleton of flinty threads built up into a framework of marvellous intricacy and grace. The lowest insect is not much more intricate, centralised, or controlled than many a worm of the sea-shore, but the ant or the bee is a very complex self-controlled organism. More exact, therefore, than any linear series, is the image of a tree with branches springing from different levels, each branch again bearing twigs some of which rise higher than the base of the branch above. A perfect scheme of this sort might not only express the facts of structure, it might also express our notions of the blood-relationships of animals and the way in which we believe that different forms have arisen.

But the wealth of form is less varied than at first sight appears. There is great wealth, but the coinage is very uniform. Our first impression is one of manifold variety; but that gives place to one of marvellous plasticity when we see how structures apparently quite different are reducible to the same general plan. Thus, as the poet Goethe first clearly showed, the seed-leaves, root-leaves, stem-leaves, and even the parts of the flower—sepals, petals, stamens, and carpels, are in reality all leaves or appendages more or less modified for diverse work. The mouth-parts of a lobster are masticating legs, and a bird's wing is a modified arm. The old naturalists were so far right in insisting on the fact of a few great types. Nature, Lamarck said, is never brusque; nor is she inventive so much as adaptive.

4. **Wealth of Numbers.**—Large numbers are so unthinkable, and accuracy in census-taking is so difficult, that we need say little as to the number of different animals. The census includes far over a million living species—a total so vast that, so far as our power of realising it is concerned, it is hardly affected when we admit that more than half are insects. To these recorded myriads, moreover, many newly-discovered forms are added every year—now by the individual workers who with fresh eye or improved microscope find in wayside pond or shore pool some new thing, or again by great enterprises like the *Challenger* expedition. Exploring naturalists like Wallace and Semper return from tropical countries enriched with new animals from the dense forests or warm seas. Zoological Stations, notably that of Naples, are "register-houses" for the fauna of the neighbouring sea, not merely as to number and form, but in many cases taking account of life and history as well. Nor can we forget the stupendous roll of the extinct, to which the zoological historians continue to add as they disentomb primitive mammals, toothed birds, giant reptiles, huge amphibians, armoured fishes, gigantic cuttles, and a vast multitude of strange forms, the like of which no longer live. The length of the *Zoological Record*, in which the literature and discoveries of each year are chronicled, the portentous size of a volume which professes to discuss with some completeness even a single sub-class, the number of special departments into which the science of zoology is divided, suggest the vast wealth of numbers at first sight so bewildering. More than two thousand years ago Aristotle recorded a total of about 500 forms, but more new species may be described in a single volume of the *Challenger* Reports. We speak about the number of the stars, yet more than one

family of insects is credited with including as many different species as there are stars to count on a clear night. But far better than any literary attempt to estimate the numerical wealth of life is some practical observation, some attempted enumeration of the inmates of your aquarium, of the tenants of some pool, or of the visitors to some meadow. The naturalist as well as the poet spoke when Goethe celebrated Nature's wealth: " In floods of life, in a storm of activity, she moves and works above and beneath, working and weaving, an endless motion, birth and death, an infinite ocean, a changeful web, a glowing life; she plies at the roaring loom of time and weaves a living garment for God."

5. **Wealth of Beauty.**—To many, however, animal life is impressive not so much because of its amazing variety and numerical greatness, nor because of its intellectual suggestiveness and practical utility, but chiefly on account of its beauty. This is to be seen and felt rather than described or talked about.

The beauty of animals, in which we all delight, is usually in form, or in colour, or in movement. Especially in the simplest animals, the beauty of form is often comparable to that of crystals; witness the marvellous architecture in flint and lime exhibited by the marine Protozoa, whose empty shells form the ooze of the great depths. In higher animals also an almost crystalline exactness of symmetry is often apparent, but we find more frequent illustration of graceful curves in form and feature, resulting in part from strenuous and healthful exercise, which moulds the body into beauty.

Not a little of the colour of animals is due to the physical nature of the skin, which is often iridescent; much, on the other hand, is due to the possession of pigments, which may either be of the nature of reserve-products, and then equivalent, let us say, to jewels, or of the nature of waste-products, and thus a literal "beauty for ashes." It is often supposed that plants excel animals in colour, but alike in the number and variety of pigments the reverse is true. Then as to movement, how much there is to admire; the birds soaring, hovering, gliding, and diving; the monkey's gymnastics; the bat's arbitrary evolutions; the grace of the fleet stag; the dolphin gamboling in the waves; the lithe lizards which flash across the path and are gone, and the snake flowing like a silver river; the buoyant swimming of fishes and all manner of aquatic animals; the lobster darting backwards with a powerful tail-stroke across the pool; the butterflies flitting like sunbeams among the flowers. But

FIG. 3.—Humming-birds (*Florisuga mellivora*) visiting flowers. (From Belt.)

are not all the delights of form and colour and movement expressed in the songs of the birds in spring?

I am quite willing to allow that this beauty is in one sense a relative quality, varying with the surroundings and education, and even ancestral history, of those who appreciate it. A flower which seems beautiful to a bee may be unattractive to a bird, a bird may choose her mate for qualities by no means winsome to human eyes, and a dog may howl painfully at our sweet music. We call the apple-blossom and the butterfly's wings beautiful, partly because the rays of light, borne from them to our eyes, cause a pleasantly harmonious activity in our brains, partly because this awakens reminiscences of past pleasant experiences, partly for subtler reasons. Still, all healthy organisms are harmonious in form, and seldom if ever are their colours out of tone with their surroundings or with each other,—a fact which suggests the truth of the Platonic conception that a living creature is harmonious because it is possessed by a single soul, the realisation of a single idea.

The plants which seem to many eyes to have least beauty are those which have been deformed or discoloured by cultivation, or taken altogether out of their natural setting; the only ugly animals are the products of domestication and human interference on the one hand, or of disease on the other; and the ugliest things are what may be called the excretions of civilisation, which are certainly not beauty for ashes, but productions by which the hues and colours of nature have been destroyed or smothered, where the natural harmony has been forcibly put out of tune—in short, where a vicious taste has insisted on becoming inventive.

CHAPTER II

THE WEB OF LIFE

1. Dependence upon Surroundings—2. Inter-relations of Plants and Animals—3. Relation of Animals to the Earth—4. Nutritive Relations—5. More Complex Interactions

IN the filmy web of the spider, threads delicate but firm bind part to part, so that the whole system is made one. The quivering fly entangled in a corner betrays itself throughout the web; often it is felt rather than seen by the lurking spinner. So in the substantial fabric of the world part is bound to part. In wind and weather, or in the business of our life, we are daily made aware of results whose first conditions are remote, and chains of influence not difficult to demonstrate link man to beast, and flower to insect. The more we know of our surroundings, the more we realise the fact that nature is a vast system of linkages, that isolation is impossible.

1. **Dependence upon Surroundings.**—Every living body is built up of various arrangements of at least twelve "elements," viz. Oxygen, Hydrogen, Carbon, Nitrogen, Chlorine, Phosphorus, Sulphur, Magnesium, Calcium, Potassium, Sodium, and Iron. All these elements are spread throughout the whole world. By the magic touch of life they are built up into substances of great complexity and instability, substances very sensitive to impulses from, or changes in, their surroundings. It may be that living matter differs from dead matter in no other way than this. The

varied forms of life crystallise out of their amorphous beginnings in a manner that we conceive to be analogous to the growth of a crystal within its solution. Further, we do not believe in a "vital force." The movements of living things are, like the movements of all matter, the expression of the world's energy, and illustrate the same laws. But to these matters we shall return in another chapter.

Interesting, because of its sharply defined and far-reaching significance, and because the essential mass is so nearly infinitesimal, is the part played by iron in the story of life. For food-supply we are dependent upon animals and plants, and ultimately upon plants. But these cannot produce their valuable food-stuffs without the green colouring-matter in their leaves, by help of which they are able to utilise the energy of sunshine and the carbonic acid gas of the air. But this important green pigment (though itself perhaps free from any iron) cannot be formed in the plant unless there be, as there almost always is, some iron in the soil. Thus our whole life is based on iron. And all our supplies of energy, our powers of doing work either with our own hands and brains, or by the use of animals, or through the application of steam, are traceable—if we follow them far enough—to the sun, which is thus the source of the energy in all creatures.

2. **Inter-relations of Plants and Animals.**—We often hear of the "balance of nature," a phrase of wide application, but very generally used to describe the mutual dependence of plants and animals. Every one will allow that most animals are more active than most plants, that the life of the former is on an average more intense and rapid than that of the latter. For all typical plants the materials and conditions of nutrition are found in water and salts absorbed by the roots, in carbonic acid gas absorbed by the leaves from the air, and in the energy of the sunlight which shines on the living matter through a screen of green pigment. Plants feed on very simple substances, at a low chemical level, and their most characteristic transformation of energy is that by which the kinetic energy of the sunlight is changed into the potential

energy of the complex stuffs which animals eat or which we use as fuel. But animals feed on plants or on creatures like themselves, and are thus saved the expense of building up food-stuffs from crude materials. Their most characteristic transformation of energy is that by which the power of complex chemical substances is used in locomotion and work. In so working, and eventually in dying, they form waste-products—water and carbonic acid, ammonia and nitrates, and so on—which may be again utilised by plants.

How often is the inaccurate statement repeated "that animals take in oxygen and give out carbonic acid, whereas plants take in carbonic acid and give out oxygen"! This is most misleading. It contrasts two entirely distinct processes—a breathing process in the animal with a feeding process in the plant. The edge is at once taken off the contrast when the student realises that plants and animals being both (though not equally) alive, must alike breathe. As they live the living matter of both is oxidised, like the fat of a burning candle; in plant, in animal, in candle, oxygen passes in, as a condition of life or combustion, and carbonic acid gas passes out as a waste-product. Herein there is no difference except in degree between plant and animal. Each lives, and must therefore breathe. But the living of plants is less intense, therefore the breathing process is less marked. Moreover, in sunlight the respiration is disguised by an exactly reverse process— peculiar to plants—the feeding already noticed, by which carbonic acid gas is absorbed, its carbon retained, and part of its oxygen liberated.

There is an old-fashioned experiment which illustrates the "balance of nature." In a glass globe, half-filled with water, are placed some minute water-plants and water-animals. The vessel is then sealed. As both the plants and the animals are absorbing oxygen and liberating carbonic acid gas, it seems as if the little living world enclosed in the globe would soon end in death. But, as we have seen, the plants are able in sunlight to absorb carbonic acid and liberate oxygen, and if present in sufficient numbers will compensate both for their own breathing and for that of animals. Thus the result within the globe need not be suffocation, but harmonious prosperity. If the minute animals ate up all the plants, they would themselves die for lack of oxygen before they had eaten up one another, while if the plants smothered all the animals they would also in turn die away. Some such contingency is apt to spoil the experiment, the end of which may be a vessel of putrid water tenanted for a long time by the very simple colourless plants known as Bacteria, and at last not even by them. Nevertheless the "vivarium" experiment is both theoretically and practically possible. Now in nature there is, indeed, no closed vivarium, for there is no isolation and there is open air, and it is an exaggeration to talk as if our life were dependent on there being a proportionate number of plants and animals in the neighbourhood. Yet the "balance of nature" is a general fact of much importance, though the economical relations of part to part over a wide area are neither rigid nor precise.

We have just mentioned the very simple plants called Bacteria. Like moulds or fungi, they depend upon other organisms for their food, being without the green colouring stuff so important in the life of most plants. These very minute Bacteria are almost omnipresent; in weakly animals —and sometimes in strong ones too—they thrive and multiply and cause death. They are our deadliest foes, but we should get rid of them more easily if we had greater love of sunlight, for this is their most potent, as well as most economical antagonist. But it is not to point out the obvious fact that a Bacterium may kill a king that we have here spoken of this class of plants; it is to acknowledge their beneficence. They are the great cleansers of the world. Animals die, and Bacteria convert their corpses into simple substances, restoring to the soil what the plants, on which the animals fed, originally absorbed through their roots. Bacteria thus complete a wide circle; they unite dead animal and living plant. For though many a plant thrives quite independently of animals on the raw materials of earth and air, others are demonstrably raising the ashes of animals into a new life. A strange partnership between Bacteria on the one hand and leguminous and cereal plants on the other has recently been discovered. There seems much likelihood that with some plants of the orders just named Bacteria live in normal partnership. The legumes and cereals in question do not thrive well without their guests, nay more, it seems as if the Bacteria are able to make the free nitrogen of the air available for their hosts.

3. **Relation of Animals to the Earth.**—Bacteria are extremely minute organisms, however, and stories of their industry are apt to sound unreal. But this cannot be said of earthworms. For these can be readily seen and watched, and their trails across the damp footpath, or their castings on the grass of lawn and meadow, are familiar to us all. They are distributed, in some form or other, over most regions of the globe; and an idea of their abundance may be gained by making a nocturnal expedition with a lantern to any convenient green plot, where they may be seen in great numbers, some crawling about, others, with their tails in their holes, making slow circuits in search of leaves and vegetable débris. Darwin estimated that there are on an average 53,000 earthworms in an acre of garden ground, that 10 tons of soil per acre pass annually through their bodies, and that they bring up mould to the surface at the rate of 3 inches thickness in fifteen years. Hensen found in his garden 64 large worm-holes in $14\frac{1}{2}$ square feet, and estimated the weight of the daily castings at about 2 cwts. in two and a half acres. In the open fields, however, it seems to be only about half as much. But whether we take Darwin's estimate that the earthworms of England pass annually through their bodies about 320,000,000 tons of earth, or the more moderate calculations of Hensen, or our own observations in the garden, we must allow that the soil-making and soil-improving work of these animals is momentous.

In Yorubaland, on the West African coast, earthworms (*Siphonogaster*) somewhat different from the common *Lumbricus* are exceedingly numerous. From two separate square feet of land chosen at random, Mr. Alvan Millson collected the worm-casts of a season and found that they weighed when dry 10¾ lbs. At this rate about 62,233 tons of subsoil would be brought in a year to the surface of each square mile, and it is also calculated that every particle of earth to the depth of two feet is brought to the surface once in 27 years. We do not wonder that the district is fertile and healthy.

Devouring the earth as they make their holes, which are often 4 or even 6 feet deep; bruising the particles in their gizzards, and thus liberating the minute elements of the soil; burying leaves and devouring them at leisure; preparing the way by their burrowing for plant roots and rain-drops, and gradually covering the surface with their castings, worms have, in the history of the habitable earth, been most important factors in progress. Ploughers before the plough, they have made the earth fruitful. It is fair, however, to acknowledge that vegetable mould sometimes forms independently of earthworms, that some other animals which burrow or which devour dead plants must also help in the process, and that the constant rain of atmospheric dust, as Richthofen has especially noted, must not be overlooked.

In 1777, Gilbert White wrote thus of the earthworms—

"The most insignificant insects and reptiles are of much more consequence and have much more influence in the economy of nature than the incurious are aware of. . . . Earthworms, though in appearance a small and despicable link in the chain of Nature, yet, if lost, would make a lamentable chasm. . . . Worms seem to be the great promoters of vegetation, which would proceed but lamely without them, by boring, perforating, and loosening the soil, and rendering it pervious to rains and the fibres of plants; by drawing straws and stalks of leaves and twigs into it; and, most of all, by throwing up such infinite numbers of lumps of earth called wormcasts, which, being their excrement, is a fine manure for grain and grass. Worms probably provide new soil for hills and slopes where the rain washes the earth away; and they affect slopes probably to avoid being flooded. . . . The earth without worms would soon become cold, hard-bound, and void of fermentation, and consequently sterile. . . . These hints we think proper to throw out, in order to set the inquisitive and discerning to work. A good monograph of worms would afford much entertainment and information at the same time, and would open a large and new field in natural history."

After a while the discerning did go to work, and Hensen published an important memoir in 1877, while Darwin's "good monograph" on the formation of vegetable mould appeared after about thirty years' observation in 1881 ; and now we all say with him, "It may be doubted whether there are many other animals which have played so important a part in the history of the world as have these lowly-organised creatures."

Prof. Drummond, while admitting the supreme importance of the work of earthworms, eloquently pleads the claims of the Termite or White Ant as an agricultural agent. This insect, which dwelt upon the earth long before the true ants, is abundant in many countries, and notably in Tropical Africa. It ravages dead wood with great rapidity. "If a man lay down to sleep with a wooden leg, it would be a heap of sawdust in the morning," while houses and decaying forest trees, furniture and fences, fall under the jaws of the hungry Termites. These fell workers are blind and live underground ; for fear of their enemies they dare not show face, and yet without coming out of their ground they cannot live.

"How do they solve the difficulty? They take the ground out along with them. I have seen white ants working on the top of a high tree, and yet they were underground. They took up some of the ground with them to the tree-top. They construct tunnels which run from beneath the soil up the sides of trees and posts ; grain after grain is carried from beneath and mortared with a sticky secretion into a reddish sandpaper-like tube ; this is rapidly extended to a great height—even of 30 feet from the ground—till some dead branch is reached. Now as many trees in a forest are thus plastered with tunnels, and as there are besides elaborate subterranean galleries and huge obelisk-like ant-hills, sometimes 10-15 feet high, it must be granted that the Termites, like the earthworms, keep the soil circulating. The earth-tubes crumble to dust, which is scattered by the wind ; the rains lash the forests and soils with fury and wash off the loosened grains to swell the alluvium of a distant valley."

The influences of plants and animals on the earth are manifold. The sea-weeds cling around the shores and lessen the shock of the breakers. The lichens eat slowly into the stones, sending their fine threads beneath the surface as thickly sometimes " as grass-roots in a meadow-land," so that the skin of the rock is gradually weathered away. On the moor the mosses form huge sponges, which mitigate floods and keep the streams flowing in days of drought. Many little plants smooth away the wrinkles on the earth's face, and adorn her with jewels; others have caught and stored the sunshine, hidden its power in strange guise in the earth, and our hearths with their smouldering peat or glowing coal are warmed by the sunlight of ancient summers. The grass which began to grow in comparatively modern (*i.e.* Tertiary) times has made the earth a fit home for flocks and herds, and protects it like a garment; the forests affect the rainfall and temper the climate, besides sheltering multitudes of living things, to some of whom every blow of the axe is a death-knell. Indeed, no plant from Bacterium to oak tree either lives or dies to itself, or is without its influence on earth and beast and man.

There are many animals besides worms which influence the earth by no means slightly. Thus, to take the minus side of the account first, we see the crayfish and their enemies the water-voles burrowing by the river banks and doing no little damage to the land, assisting in that process by which the surface of continents tends gradually to diminish. So along the shores in the harder substance of the rocks there are numerous borers, like the Pholad bivalves, whose work of disintegration is individually slight, but in sum-total great. More conspicuous, however, is the work of the beavers, who, by cutting down trees, building dams, digging canals, have cleared away forests, flooded low grounds, and changed the aspect of even large tracts of country. Then, as every one knows, there are injurious insects innumerable, whose influence on vegetation, on other animals, and on the prosperity of nations, is often disastrously great.

But, on the other hand, animals cease not to pay their

filial debts to mother earth. We see life rising like a mist in the sea, lowly creatures living in shells that are like mosques of lime and flint, dying in due season, and sinking gently to find a grave in the ooze. We see the submarine volcano top, which did not reach the surface of the ocean, slowly raised by the rainfall of countless small shells. Inch by inch for myriads of years, the snow-drift of dead shells forms a patient preparation for the coral island. The tiniest, hardly bigger than the wind-blown dust, form when added together the strongest foundation in the world. The vast whale skeleton falls, but melts away till only the ear-bones are left. Of the ruthless gristly shark nothing stays but teeth. The sea-butterflies (Pteropods), with their frail shells, are mightier than these, and perhaps the microscopic atomies are strongest of all. The pile slowly rises, and the exquisite fragments are cemented into a stable foundation for the future city of corals.

At length, when the height at which they can live is reached, coral germs moor themselves to the sides of the raised mound, and begin a new life on the shoulders of death. They spread in brightly coloured festoons, and have often been likened to flowers. The waste salts of their living perhaps unite with the gypsum of the sea-water, at any rate in some way the originally soft young corals acquire strong shells of carbonate of lime. Sluggish creatures they, living in calcareous castles of indolence! In silence they spread, and crowd and smother one another in a struggle for standing-room. The dead forms, partly dissolved and cemented, become a broad and solid base for higher and higher growth. At a certain height the action of the breakers begins, great severed masses are piled up or roll down the sloping sides. Clear daylight at last is reached, the mound rises above the water. The foundations are ever broadened, as vigorously out-growing masses succumb to the brunt of the waves and tumble downwards. Within the surface-circle weathering makes a soil, and birds resting there with weary wings, or perhaps dying, leave many seeds of plants—the beginnings of another life. The waves cast up forms of dormant life which have floated from afar, and a terrestrial fauna and flora begin. It is a strange and beautiful story, dead shells of the tenderest beauty on the rugged shoulders of the volcano; corals like meadow flowers on the graveyard of the ooze; at last plants and trees, the hum of insects and the song of birds, over the coral island.

4. **Nutritive Relations.**—What we may call "nutritive chains" connect many forms of life—higher animals feeding upon lower through long series, the records of which sound like the story of "The House that Jack built." On land and on the shore these series are usually short, for plants are abundant, and the carnivores feed on the vegetarians. In the open sea, where there is less vegetation, and in the great depths, where there is none, carnivore preys upon carnivore throughout long series—fish feeds upon fish, fish upon crustacean, crustacean upon worm, worm on débris. Disease or disaster in one link affects the whole chain. A parasitic insect, we are told, has killed off the wild horses and cattle in Paraguay, thereby influencing the vegetation, thereby the insects, thereby the birds. Birds of prey and small mammals—so-called "vermin"—are killed off in order to preserve the grouse, yet this interference seems in part to defeat itself by making the survival of weak and diseased birds unnaturally easy, and epidemics of grouse-disease on this account the more prevalent. A craze of vanity or gluttony leads men to slaughter small insect-eating birds, but the punishment falls—unluckily on the wrong shoulders—when the insects which the birds would have kept down increase in unchecked numbers, and destroy the crops of grain and fruit. In a fuel-famine men have sometimes been forced to cut down the woods which clothe the sides of a valley, an action repented of when the rain-storms wash the hills to skeletons, when the valley is flooded and the local climate altered, and when the birds robbed of their shelter leave the district to be ravaged by caterpillar and fly. American entomologists have proved that the ravages of destructive insects may be checked by importing and fostering their natural enemies, and on the other hand, the sparrows which have established themselves in the States

have in some districts driven away the titmice and thus favoured the survival of injurious caterpillars.

5. **More Complex Interactions.**—The flowering plants and the higher insects have grown up throughout long ages together, in alternate influence and mutual perfecting. They now exhibit a notable degree of mutual dependence; the insects are adapted for sipping the nectar from the blossoms; the flowers are fitted for giving or receiving the fertilising golden dust or pollen which their visitors, often quite unconsciously, carry from plant to plant. The mouth organs of the insects have to be interpreted in relation to the flowers which they visit; while the latter show structures which may be spoken of as the "footprints" of the insects. So exact is the mutual adaptation that Darwin ventured to prophesy from the existence of a Madagascar orchid with a nectar-spur 11 inches long, that a butterfly would be found in the same locality with a suctorial proboscis long enough to drain the cup; and Forbes confirmed the prediction by discovering the insect.

As information on the relations of flowers and insects is readily attainable, and as the subject will be discussed in the volume on Botany, it is sufficient here to notice that, so far as we can infer from the history half hidden in the rocks, the floral world must have received a marked impulse when bees and other flower-visiting insects appeared; that for the successful propagation of flowering plants it is advantageous that pollen should be carried from one individual to another, in other words, that cross-fertilisation should be effected; and that, for the great majority of flowering plants, this is done through the agency of insects. How plants became bright in colour, fragrant in scent, rich in nectar, we cannot here discuss; the fact that they are so is evident, while it is also certain that insects are attracted by the colour, the scent, and the sweets. Nor can there be any hesitation in drawing the inference that the flowers which attracted insects with most success, and insects which got most out of the flowers, would, *ipso facto*, succeed better in life.

No illustration of the web of life can be better than the most familiar one, in which Darwin traced the links of influence between cats and clover. If the possible seeds in the flowers of the purple clover are to become real seeds, they must be fertilised by the golden dust or pollen from some adjacent clover plants. But as this pollen is unconsciously carried from flower to flower by the humble-bees, the proposition must be granted that the more humble-bees, the better next year's clover crop. The humble-bees, however, have their enemies in the field-mice, which lose no opportunity of destroying the combs; so that the fewer field-mice, the more humble-bees, and the better next year's clover crop. In the neighbourhood of villages, however, it is well known that the cats make as effective war on the field-mice as the latter do on the bees. So that next year's crop of purple clover is influenced by the number of humble-bees, which varies with the number of field-mice, that is to say, with the abundance of cats; or, to go a step farther, with the number of lonely ladies in the village. It should be noted, however, that according to Mr. James Sime there were abundant fertile clover crops in New Zealand before there were any humble-bees in that island. Indeed, many think that the necessity of cross-fertilisation has been exaggerated.

Not all insects, however, are welcome visitors to plants; there are unbidden guests who do harm. To their visits, however, there are often obstacles. Stiff hairs, impassably slippery or viscid stems, moats in which the intruders drown, and other structural peculiarities, whose origin may have had no reference to insects, often justify themselves by saving the plant. Even more interesting, however, is the preservation of some acacias and other shrubs by a bodyguard of ants, which, innocent themselves, ward off the attacks of the deadly leaf-cutters. In some cases the bodyguard has become almost hereditarily accustomed to the plants, and the plants to them, for they are found in constant companionship, and the plants exhibit structures which look almost as if they had been made as shelters for the ants. On some of our European trees similar little homes or *domatia* constantly occur, and shelter small insects which do no harm to the trees, but cleanse them from injurious fungi.

In many ways plants are saved from the appetite of animals. The nettle has poisonous hairs; thistles, furze, and holly are covered with spines; the hawthorn has its thorns and the rose its prickles; some have repulsive odours; others contain oils, acids, ferments, and poisons which many animals dislike; the cuckoo-pint (*Arum*) is full of little crystals which make our lips smart if we nibble a leaf. In our studies of plants we endeavour to find out what these qualities primarily mean to their possessors; here we think rather of their secondary significance as protections against animals. For though snails ravage all the plants in a district except those which are repulsive, the snails are at most only the secondary factors in the evolution of the repulsive qualities.

FIG. 4.—Acacia (*A. sphærocephala*), with hollow thorns in which ants find shelter. (After Schimper.)

The strange inter-relations between plants and animals are again illustrated by the carnivorous, generally insectivorous, plants. It is not our business to discuss the original or primary import of the pitchers of pitcher-plants or of the mobile and sensitive leaves of Venus' Fly-Trap; nowadays, at any rate, insects are attracted to them, captured by them, and used. Let us take only one case, that of the common Bladderwort (*Utricularia*). Many of the leaflets of this plant, which floats in summer in the marsh pond, are modified into little bladders, so fashioned that minute "water-fleas"—which swarm in every corner of the pool—can readily enter them, but can in no wise get out again. The small entrance is guarded by a valve or door, which opens inwards, but allows no egress. The little crustaceans are attracted by some mucilage made by the leaves, or sometimes perhaps by sheer curiosity; they enter and cannot return; they die, and their débris is absorbed by the leaf.

Again, in regard to distribution, there are numerous relations between organisms. Spiny fruits like those of Jack-run-the-hedge adhere to animals, and are borne from place to place; and minute water-plants and animals are carried from one watercourse to another on the muddy feet of birds. Darwin removed a ball of mud from the leg of a bird, and from it fourscore seeds germinated. Not a bird can fall to the ground and die without sending a throb through a wide circle.

A conception of these chains or circles of influence is important, not only for the sake of knowledge, but also as a guide in action. Thus, to take only one instance among a hundred, it may seem a far cry from a lady's toilet-table to the African slave-trade, but when we remember the ivory backs of the brushes, and how the slaves are mainly used for transporting the tusks of elephants—a doomed race—from the interior to the coast, the riddle is read, and the responsibility is obvious. Over a ploughed field in the summer morning we see the spider-webs in thousands glistening with mist-drops, and this is an emblem of the intricacy of the threads in the web of life—to be seen more and more as our eyes grow clear. Or, is not the face of nature like the surface of a gentle stream, where hundreds of dimpling circles touch and influence one another in an infinite complexity of action and reaction beyond the ken of the wisest?

CHAPTER III

THE STRUGGLE OF LIFE

1. *Nature and Extent of the Struggle*—2. *Armour and Weapons*—3. *Different Forms of Struggle*—4. *Cruelty of the Struggle*

1. **Nature and Extent of the Struggle.**—If we realise what is meant by the "web of life," the recognition of the "struggle for existence" cannot be difficult. Animals do not live in isolation, neither do they always pursue paths of peace. Nature is not like a menagerie where beast is separated from beast by iron bars, neither is it a mêlée such as would result if the bars of all the cages were at once removed. It is not a continuous Waterloo, nor yet an amiable compromise between weaklings. The truth lies between these extremes. In most places where animals abound there is struggle. This may be silent and yet decisive, real without being very cruel, or it may be full of both noise and bloodshed.

This struggle is very old; it is older than the conflicts of men, older than the ravin of tooth and claw, it is as old as life. The struggle is often very keen—often for life or death. But though few animals escape experience of the battlefield—and for some there seems no discharge from this war—we must not misinterpret nature as "a continual free-fight." One naturalist says that all nature breathes a hymn of love, but he is an optimist under sunny southern skies; another compares nature to a huge gladiatorial show with a plethora of fighters, but he speaks as a pessimist from amid the din of individualistic competition. Nature is full of struggle and fear, but the struggle is sometimes outdone by sacrifice, and the fear is sometimes cast out by love. We must be careful to remember Darwin's proviso that he used the phrase "struggle for existence" "in a large and metaphorical sense, including the dependence of one being on another, and including (which is more important) not only the life of the individual, but success in leaving progeny." He also acknowledged the importance of mutual aid, sociability, and sympathy among animals, though he did not carefully estimate the relative importance of competition on the one hand and sociability on the other. Discussing sympathy, Darwin wrote, "In however complex a manner this feeling may have originated, as it is one of high importance to all those animals which aid and defend one another, it will have been increased through natural selection; for those communities which included the greatest number of the most sympathetic members would flourish best, and rear the greatest number of offspring." I should be sorry to misrepresent the opinions of any man, but after considerable study of modern Darwinian literature, I feel bound to join in the protest which others have raised against a tendency to narrow Darwin's conception of "the struggle for existence," by exaggerating the occurrence of internecine competitive struggle. Thus Huxley says, "Life was a continuous free-fight, and beyond the limited and temporary relations of the family, the Hobbesian war of each against all was the normal state of existence." Against which Kropotkine maintains that this "view of nature has as little claim to be taken as a scientific deduction as the opposite view of Rousseau, who saw in nature but love, peace, and harmony destroyed by the accession of man." . . . "Rousseau has committed the error of excluding the beak-and-claw fight from his thoughts, and Huxley is committing the opposite error; but neither Rousseau's optimism nor Huxley's pessimism can be accepted as an impartial interpretation of nature."

2. **Armour and Weapons.**—If you doubt the reality

of the struggle, take a survey of the different classes of animals. Everywhere they brandish weapons or are fortified with armour. "The world," Diderot said, "is the abode of the strong." Even some of the simplest animals have offensive threads, prophetic of the poisonous lassoes with which jellyfish and sea-anemones are equipped. Many worms have horny jaws; crustaceans have strong pincers; many insects have stings, not to speak of mouth organs like surgical instruments; spiders give poisonous bites; snails have burglars' files; the cuttlefish have strangling suckers and parrots' beaks. Among backboned animals we recall the teeth of the shark and the sword of the swordfish, the venomous fangs of serpents, the jaws of crocodiles, the beaks and talons of birds, the horns and hoofs and canines of mammals. Now we do not say that these and a hundred other weapons were from their first appearance weapons, indeed we know that most of them were not. But they are weapons now, and just as we would conclude that there was considerable struggle in a community where every man bore a revolver, we must draw a similar inference from the offensive equipment of animals.

As to armoured beasts, we remember that shells of lime or flint occur in many of the simplest animals, that most sponges are so rich in spicules that they are too gritty to be pleasant eating, that corals are polypes within shells of lime, that many worms live in tubes, that the members of the starfish class are in varying degrees lime-clad, that crustaceans and insects are emphatically armoured animals, and that the majority of molluscs live in shells. So among backboned animals, how thoroughly bucklered were the fishes of the old red sandstone against hardly less effective teeth, how the scales of modern fishes glitter, how securely the sturgeon swims with its coat of bony mail! Amphibians are mostly weaponless and armourless, but reptiles are scaly animals *par excellence*, and the tortoise, for instance, lives in an almost impregnable citadel. Birds soar above pursuit, and mammals are swift and strong, but among the latter the armadillos have bony shields of marvellous strength, and hedgehog and porcupine have their hair hardened into spines and quills. Now we do not say that all these structures were from the first of the nature of armour, indeed they admit of other explanations, but that they serve as armour now there can be no doubt. And just as we conclude that a man would not wear a chain shirt without due reason, so we argue from the prevalence of animal armour to the reality of struggle.

For a moment let me delay to explain the two saving-clauses which I have inserted. The pincers of a crab are modified legs, the sting of a bee has probably the same origin, and it is likely that most weapons originally served some other than offensive purpose. We hear of spears becoming pruning-hooks; the reverse has sometimes been true alike of animals and of men. By sheer use a structure not originally a weapon became strong to slay; for there is a profound biological truth in the French proverb: "*A force de forger on devient forgeron.*"

And again as to armour, it is, or was, well known that a boy's hand often smitten by the "tawse" became callous as to its epidermis. Now that callousness was not a device—providential or otherwise—to save the youth from the pains of chastisement, and yet it had that effect. By bearing blows one naturally and necessarily becomes thick-skinned. Moreover, the epidermic callousness referred to might be acquired by work or play altogether apart from school discipline, though it might also be the effect of the blows. In the same way many structures which are most useful as armour may be the "mechanical" or natural results of what they afterwards help to obviate, or they may arise quite apart from their future significance.

3. **Different Forms of Struggle.**—If you ask why animals do not live at peace, I answer, *more Scottico*, Why do not we? The desires of animals conflict with those of their neighbours, hence the struggle for bread and the competition for mates. Hunger and love solve the world's problems. Mouths have to be filled, but population tends locally and temporarily to outrun the means of subsistence, and the question "which mouths"

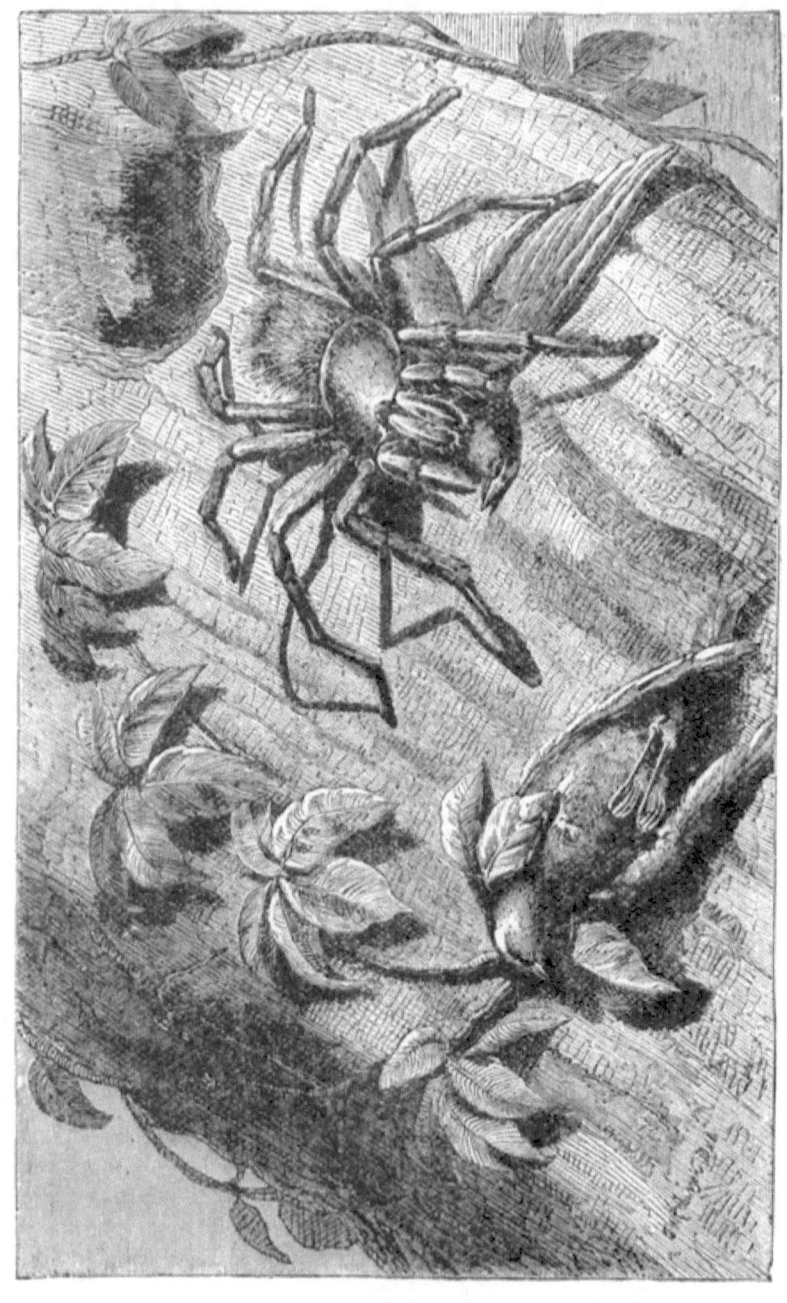

FIG. 5.—Bird-catching spider (*Mygale avicularia*) attacking finches. (From Bates.)

has to be decided—sometimes by peaceful endeavour, as in migration, sometimes with teeth clenched or ravenous. Many animals are carnivorous, and must prey upon weaker forms, which do their best to resist. Mates also have to be won, and lover may fight with lover till death is stronger than both. But these struggles for food and for mates are often strivings rather than strife, nor is a recognition of the frequent keenness and fierceness of the competition inconsistent with the recognition of mutual aid, sociability, and love. There is a third form of the struggle,—that between an animal and its changeful surroundings. This also is a struggle without strife. Fellow competitors strive for their share of the limited means of subsistence; between foes there is incessant thrust and parry; in the courtship of mates there are many disappointed and worsted suitors; over all are the shears of fate—a changeful physical environment which has no mercy.

An analysis of the various forms of struggle may be attempted as follows:

For Food	(*a*)	Between animals of the same kind which compete for similar food and other necessaries of life—*Struggle between fellows*.
	(*b*)	Between animals of different kinds, the one set striving to devour, the other set endeavouring to escape their foes, *e.g.* between carnivores and herbivores—*Struggle between foes*.
For Love	(*c*)	Between the rival suitors for desired mates—*Struggle between rivals in love*.
For Foot-hold	(*d*)	Between animals and changeful surroundings—*Struggle with fate*.

In most cases, besides the egoism or individualism, one must recognise the existence of altruism, parental love and sacrifice, mutual aid, care for others, and sociality.

Before we consider these different forms of struggle, let us notice the rapid multiplication of individuals which furnishes the material for what in "a wide and metaphorical sense" may be called a "battlefield."

A single Infusorian may be the ancestor of millions by the end of a week. A female aphis, often producing one offspring per hour for days together, might in a season be the ancestor of a progeny of atomies which would weigh down five hundred millions of stout men. "The roe of a cod contains sometimes nearly ten million eggs, and supposing each of these produced a young fish which arrived at maturity, the whole sea would immediately become a solid mass of closely packed codfish." The unchecked multiplication of a few mice or rabbits would soon leave no standing-room on earth.

But fortunately, with the exception of the Infusorians, these multiplications do not occur. We have to thank the struggle in nature, and especially the physical environment, that they do not. The fable of Mirza's bridge is continually true,—few get across.

(*a*) It is often said that the struggle between fellows of the same kind and with the same needs is keenest of all, but this is rather an assumption than an induction from facts. The widespread opinion is partly due to an *a priori* consideration of the problem, partly to that anthropomorphism which so easily besets us. We transfer to the animal world our own experience of keen competition with fellows of the same caste, and in so doing are probably unjust. Thus Mr. Grant Allen says—

"The baker does not fear the competition of the butcher in the struggle for life; it is the competition of the other bakers that sometimes inexorably crushes him out of existence. . . . In this way the great enemies of the individual herbivores are not the carnivores, but the other herbivores. . . . It is not so much the battle between the tiger and the antelope, between the wolf and the bison, between the snake and the bird, that ultimately results in natural selection or survival of the fittest, as the struggle between tiger and tiger, between bison and bison, between snake and snake, between antelope and antelope. . . . *Homo homini lupus*, says the old proverb, and so, we may add, in a wider sense, *lupus lupo lupus*, also. . . . The struggle is fierce between allied kinds, and fiercest of all between individual members of the same species."

I have quoted these sentences because they are clearly and cleverly expressed, after the manner of Grant Allen, but I do not believe that they are true statements of facts. The evidence is very unsatisfactory. In his paragraph summarised as "struggle for life most severe between individuals and varieties of the same species; often severe between species of the same genus," Darwin gave five illustrations: one species of swallow is said to have ousted another in North America, the missel-thrush has increased in Scotland at the expense of the song-thrush, the brown rat displaces the black rat, the small Asiatic cockroach drives its great congener before it, the hive-bee imported to Australia is rapidly exterminating the small, stingless native bee. But the cogency of these instances may be disputed: thus what is said about the thrushes is denied by Professor Newton. And on the other hand, we know that reindeer, beavers, lemming, buffaloes and many other animals migrate when the means of subsistence are unequal to the demands of the population, and there are other peaceful devices by which animals have discovered a way out of a situation in which a life-and-death struggle might seem inevitable. Very instructive is the fact that beavers, when too numerous in one locality, divide into two parties and migrate up and down stream. The old proverb which Grant Allen quotes, *Homo homini lupus*, appears to me a libellous inaccuracy; the extension of the libel to the animal world has certainly not been justified by careful induction. For a discussion of the alleged competition between fellows, I refer, and that with pleasure and gratitude, to Kropotkine's articles on "Mutual Aid among Animals," *Nineteenth Century*, September and November 1890.

(*b*) Of the struggle between foes differing widely in kind little need be said. It is very apparent, especially in wild countries. Carnivores prey upon herbivores, which sometimes unite in successful resistance. Birds of prey devour small mammals, and sometimes have to fight hard for their booty. Reptiles also have their battles—witness the combats between snake and mongoose. In many cases, however, carnivorous animals depend upon small fry; thus many birds feed on fishes, insects, and worms, and many fishes live on minute crustaceans. In such cases the term

FIG. 6.—Weasel attacking a grouse. (From St. John's *Wild Sports*.)

struggle must again be used "in a wide and metaphorical sense."

(*c*) In a great number of cases there is between rival males a contest for the possession of the females,—a competition in which beauty and winsomeness are sometimes as important as strength. Contrast the musical competition between rival songsters with the fierce combats of the stags.

Many animals are not monogamous, and this causes strife; a male seal, for instance, guards his harem with ferocity.

(*d*) Finally, physical nature is quite careless of life. Changes of medium, temperature, and moisture, continually occur, and the animals flee for their lives, adapt themselves to new conditions, or perish. Cataclysms are rare, but changes are common, and especially in such schools of experience as the sea-shore we may study how vicissitude has its victims or its victors.

The struggle with Fate, that is to say, with changeful surroundings, is more pleasant to contemplate than the other kinds of struggle, for at the rigid mercilessness of physical nature we shudder less than at the cruel competition between living things, and we are pleased with the devices by which animals keep their foothold against wind and weather, storm and tide, drought and cold. One illustration must suffice: drought is common, pools are dried up, the inhabitants are left to perish. But often the organism draws itself together, sweats off a protective sheath, which is not a shroud, and waits until the rain refreshes the pools. Not the simplest animals only, but some of comparatively high degree, are thus able to survive desiccation. The simplest animals encyst, and may be blown about by the wind, but they rest where moisture moors them, and are soon as lively as ever. Leaping a long way upwards, we find that the mud-fish (*Protopterus*) can be transported from Africa to Northern Europe, dormant, yet alive, within its ball of clay. We do not believe in toads appearing out of marble mantelpieces, and a palæontologist will but smile if you tell him of a frog which emerged from an intact piece of old red sandstone, but amphibians may remain for a long time dormant either in the mud of their native pools or in some out-of-the-way chink whither they had wandered in their fearsome youth.

A shop which had once been used in the preparation of bone-dust was after prolonged emptiness reinstated in a new capacity. But it was soon fearfully infested with mites (*Glyciphagus*), which had been harboured in crevices in a strange state of dry dormancy. Every mite had in a sense died, but remnant cells in the body of each had clubbed together in a life-preserving union so effective that a return of prosperity was followed by a reconstitution of mites and by a plague of them. Of course great caution must be exercised with regard to all such stories, as well as in regard to the toads within stones. Of common little animals known as Rotifers, it is often said, and sometimes rightly, that they can survive prolonged desiccation. In a small pool on the top of a granite block, there flourished a family of these Rotifers. Now this little pool was periodically swept dry by the wind, and in the hollow there remained only a scum of dust. But when the rain returned and filled the pool, there were the Rotifers as lively as ever. What inference was more natural than that the Rotifers survived the desiccation, and lay dormant till moisture returned? But Professor Zacharias thought he would like to observe the actual revivification, and taking some of the dusty scum home, placed it under his microscope on a moist slide, and waited results. There were the corpses of the Rotifers plain enough, but they did *not* revive even in abundant moisture. What was the explanation? The *eggs* of these Rotifers survived, they developed rapidly, they reinstated the family. And of course it is much easier to understand how single cells, as eggs are, could survive being dried up, while their much more complex parents perished. I do not suggest that no Rotifers can survive desiccation, it is certain that some do; but the story I have told shows the need of caution. There is no doubt, moreover, that certain simple "worms," known as "paste-eels," "vinegar-eels," etc., from their frequent occurrence in such substances, can survive desiccation for many years. Repeated experiments have shown that they can lie dormant for as long as, but not longer than, fourteen years! and it is interesting to notice that the more prolonged the period of desiccation has been, the longer do these threadworms take to revive after moisture has been supplied. It seems as if the life retreated further and further, till at length it may retreat beyond recall. In regard to plants there are many similar facts, for though accounts of the germination of seeds from the mummies of the pyramids, or from the graves of the Incas, are far from satisfactory, there is no doubt that seeds of cereals and leguminous plants may retain their life in a dormant state for years, or even for tens of years.

But desiccation is only one illustration out of a score of the manner in which animals keep their foothold against fate. I need hardly say that they are often unsuccessful; the individual has often fearful odds against it. How many winged seeds out of a thousand reach a fit resting-place where they may germinate? Professor Möbius says that out of a million oyster embryos only one individual grows up, a mortality due to untoward currents and surroundings, as well as to hungry mouths. Yet the average number of thistles and oysters tends to continue, "So careful of the type she seems, so careless of the single life." Yet though the average usually remains constant, there is no use trying to ignore, what Richard Jefferies sometimes exaggerated, that the physical fates are cruel to life. But how much wisdom have they drilled into us?

> " For life is not as idle ore,
> But iron dug from central gloom,
> And heated hot with burning fears,
> And dipt in baths of hissing tears,
> And battered by the shocks of doom
> To shape and use."

4. **Cruelty of the Struggle.**—Opinions differ much as to the cruelty of the "struggle for existence," and the question is one of interest and importance. Alfred Russel Wallace and others try to persuade us that our conception of the "cruelty of nature" is an anthropomorphism; that, like Balbus, animals do not fear death; that the rabbit rather enjoys a run before the fox; that thrilling pain soon brings its own anæsthetic; that violent death has its pleasures, and starvation its excitement. Mr. Wallace, who speaks with the authority of long and wide experience, enters a vigorous protest against Professor Huxley's description of the myriads of generations of

herbivorous animals "which have been tormented and devoured by carnivores"; of both alike "subject to all the miseries incidental to old age, disease, and over-multiplication"; of the "more or less enduring suffering" which is the meed of both vanquished and victor; of the whole creation groaning in pain. "There is good reason to believe," says Mr. Wallace, "that the supposed torments and miseries of animals have little real existence, but are the reflection of the imagined sensations of cultivated men and women in similar circumstances; and that the amount of actual suffering caused by the struggle for existence among animals is altogether insignificant." "Animals are spared from the pain of anticipating death; violent deaths, if not too prolonged, are painless and easy; neither do those which die of cold or hunger suffer much; the popular idea of the struggle for existence entailing misery and pain on the animal world is the very reverse of the truth." He concludes by quoting the conclusion of Darwin's chapter on the struggle for existence: "When we reflect on this struggle, we may console ourselves with the full belief that the war of nature is not incessant, that no fear is felt, that death is generally prompt, and that the vigorous, the healthy, and the happy survive and multiply." Yet it was Darwin who confessed that he found in the world "too much misery."

We have so little security in appreciating the real life—the mental and physical pain or happiness—of animals, that there is apt to be exaggeration on both sides, according as a pessimistic or an optimistic mood predominates. I therefore leave it to be settled by your own observation whether hunted and captured, dying and starving, maimed and half-frozen animals have to endure "an altogether insignificant amount of actual suffering in the struggle for existence."

But I think we must admit that there is much truth in what Mr. Wallace urges. Moreover, the term cruelty can hardly be used with accuracy when the involved infliction of pain is necessary. In many cases the carnivores are less "cruel" to their victims than we are to our domesticated animals. We must also remember that the "struggle for existence" is often applicable only in its "wide and metaphorical sense." And it is fair to balance the happiness and mutual helpfulness of animals against the pain and deathful competition which undoubtedly exist.

What we must protest against is that one-sided interpretation according to which individualistic competition is nature's sole method of progress. We are told that animals have got on by their struggle for individual ends; that they have made progress on the corpses of their fellows, by a "blood and iron" competition in which each looks out for himself, and extinction besets the hindmost. To those who accept this interpretation the means employed seem justified by the results attained. But it is only in after-dinner talk that we can slur over whatever there is of pain and cruelty, overcrowding and starvation, hate and individualism, by saying complacently that they are justified in us their children; that we can rest satisfied that what has been called "a scheme of salvation for the elect by the damnation of the vast majority" is a true statement of the facts; that we can seriously accept a one-sided account of nature's regime as a justification of our own ethical and economic practice.

The conclusions, which I shall afterwards seek to substantiate, are, that the struggle for existence, with its associated natural selection, often involves cruelty, but certainly does not always do so; that joy and happiness, helpfulness and co-operation, love and sacrifice, are also facts of nature, that they also are justified by natural selection; that the precise nature of the means employed and ends attained must be carefully considered when we seek from the records of animal evolution support or justification for human conduct; and that the tragic chapters in the history of animals (and of men) must be philosophically considered in such light as we can gather from what we know of the whole book.

CHAPTER IV

SHIFTS FOR A LIVING

1. *Insulation*—2. *Concealment*—3. *Parasitism*—4. *General Resemblance to Surroundings*—5. *Variable Colouring*—6. *Rapid Change of Colour*—7. *Special Protective Resemblance*—8. *Warning Colours*—9. *Mimicry*—10. *Masking*—11. *Combination of Advantageous Qualities*—12. *Surrender of Parts*

GRANTING the struggle with fellows, foes, and fate, we are led by force of sympathy as well as of logic to think of the shifts for a living which tend to be evolved in such conditions, and also of some other ways by which animals escape from the intensity of the struggle.

1. **Insulation.**—Some animals have got out of the struggle through no merit of their own, but as the result of geological changes which have insulated them from their enemies. Thus, in Cretaceous times probably, the marsupials which inhabited the Australasian region were insulated. In that region they were then the only representatives of Mammalia, and so, excepting the "native dog," some rodents and bats, and more modern imports, they still continue to be. By their insulation they were saved from that contest with stronger mammals in which all the marsupials left on the other continents were exterminated, with the exception of the opossums, which hide in American forests. A similar geological insulation accounts for the large number of lemurs in the island of Madagascar.

2. **Concealment.**—A change of habitat and mode of life is often as significant for animals as it is for men. It is easy to understand how mammals which passed from terrestrial to more or less aquatic life, for instance beaver and polar bear, seals, and perhaps whales, would enjoy a period of relative immunity after the awkward time of transition was over. So, too, many must have passed from the battlefield of the sea-shore to relative peace on land or in the deep-sea. In a change from open air to underground life, illustrated for instance in the mole, many animals have sought and found safety, and the change seems even now in progress, as in the New Zealand parrot *Stringops*, which, having lost the power of flight, has taken to burrowing. Similarly the power of flight must have helped insects, some ancient saurians, and birds out of many a scrape, though it cannot be doubted that this transition, and also that from diurnal to nocturnal habits, often brought only a temporary relief.

3. **Parasitism.**—From the simple Protozoa up to the beginning of the backboned series, we find illustrations of animals which have taken to a thievish existence as unbidden guests in or on other organisms. Flukes, tapeworms, and some other "worms," many crustaceans, insects, and mites, are the most notable. Few animals are free from some kind of parasite. There are various grades of parasitism; there are temporary and permanent, external and internal, very degenerate, and very slightly affected parasites. Sometimes the adults are parasitic while the young are free-living, sometimes the reverse is true; sometimes the parasite completes its life in one host, often it reaches maturity only after the host in which its youth has been passed is devoured by another. In many cases the habit was probably begun by the females, which seek shelter during the period of egg-laying; in not a few crustaceans and insects the females alone are parasitic. Most often, in all probability, hunger and the search for shelter led to the establishment of the thievish habit. Now, the advantages gained by a thoroughgoing parasite are great—safety, warmth, abundant food, in short, "complete material well-being." But

there is another aspect of the case. Parasitism tends to be followed by degeneration — of appendages, food-canal, sense-organs, nervous system, and other structures, the possession and use of which make life worth living. Moreover, though the reproductive system never degenerates, the odds are often many against an embryo reaching a fit host or attaining maturity. Thus Leuckart calculates that a tapeworm embryo has only about 1 chance in 83,000,000 of becoming a tapeworm, and one cannot be sorry that its chance is not greater. In illustration of the degeneration which is often associated with parasitism, and varies as the habit is more or less predominant, take the case of *Sacculina*—a crustacean usually ranked along with barnacles and acorn-shells. It begins its life as a minute free "nauplius," with three pairs of appendages, a short food-canal, an eye, a small brain, and some other structures characteristic of many young crustaceans. In spite of this promiseful beginning, the young *Sacculina* becomes a parasite, first within the body, and finally under the tail, of a crab. Attached by absorptive suckers to its host, and often doing no slight damage, it degenerates into an oval sac, almost without trace of its former structure, with reproductive system alone well developed. Yet the degeneration is seldom so great as this, and it is fair to state that many parasites, especially those which remain as external hangers-on, seem to be but slightly affected by their lazy thievish habit; nor can it be denied that most are well adapted to the conditions of their life. But on the whole the parasitic life tends to degeneration, and is unprogressive. Meredith writes of Nature's sifting—

> "Behold the life of ease, it drifts.
> The sharpened life commands its course:
> She winnows, winnows roughly, sifts,
> To dip her chosen in her source.
> Contention is the vital force
> Whence pluck they brain, her prize of gifts."

4. General Resemblance to Surroundings. — Many transparent and translucent blue animals are hardly visible in the sea; white animals, such as the polar bear, the arctic fox, and the ptarmigan in its winter plumage, are inconspicuous upon the snow; green animals, such as insects, tree-frogs, lizards, and snakes, hide among the leaves and herbage ; tawny animals harmonise with sandy soil ; and the hare escapes detection among the clods. So do spotted animals such as snakes and leopards live unseen in the interrupted light of the forest, and the striped tiger is lost in the jungle. Even the eggs of birds are often well suited to the surroundings in which they are laid. There can be no doubt that this resemblance between the colour of an animal and that of its surroundings is sometimes of protective and also aggressive value in the struggle for existence, and where this is the case, natural selection would foster it, favouring with success those variations which were best adapted, and eliminating those which were conspicuous.

But there are many instances of resemblance to surroundings which are hard to explain. Thus Dr. A. Seitz describes a restricted area of woodland in South Brazil, where the great majority of the insects were blue, although but a few miles off a red colour was dominant. He maintains that the facts cannot in this case be explained as due either to general protective resemblance or to mimicry.

I have reduced what I had written in illustration of advantageous colouring of various kinds, because this exceedingly interesting subject has been treated in a readily available volume by one who has devoted much time and skill to its elucidation. Mr. E. B. Poulton's *Colours of Animals* (International Science Series, London, 1890) is a fascinating volume, for which all interested in these aspects of natural history must be grateful. With this a forthcoming work (*Animal Coloration*, London, 1892) by Mr. F. E. Beddard should be compared.

5. **Variable Colouring.**—Some animals, such as the ptarmigan and the mountain-hare, become white in winter, and are thereby safer and warmer. In some cases it is certain that the pigmented feathers and hairs *become* white, in other cases the old feathers and hairs drop

E

off and are *replaced* by white ones; sometimes the whiteness is the result of both these processes. It is directly due to the formation of gas bubbles inside the hairs or feathers in sufficient quantity to antagonise the effect of any pigment that may be present, but in the case of new growths it is not likely that any pigment is formed. In some cases, *e.g.* Ross's lemming and the American hare (*Lepus americanus*), it has been clearly shown that the change is due to the cold. It is likely that this acts somewhat indirectly upon the skin through the nervous system. We may therefore regard the change as a variation due to the environment, and it is at least possible that the permanent whiteness of some northern animals, *e.g.* the polar bear, is an acquired character of similar origin. There are many objections to the theory that the winter whiteness of arctic animals arose by the accumulation of small variations in individuals which, being slightly whiter than their neighbours, became dominant by natural selection, though there can be no doubt that the whiteness, however it arose, would be conserved like other advantageous variations.

To several naturalists, but above all to Mr. Poulton, we are indebted for much precise information in regard to the variable colouring of many caterpillars and chrysalides. These adjust their colours to those of the surroundings, and even the cocoons are sometimes harmoniously coloured. There is no doubt that the variable colouring often has protective value. Mr. Poulton experimented with the caterpillars of the peacock butterfly (*Vanessa io*), small tortoise-shell (*Vanessa urticæ*), garden whites (*Pieris brassicæ* and *Pieris rapæ*), and many others. Caterpillars of the small tortoise-shell in black surroundings tend to become darker as pupæ; in a white environment the pupæ are lighter; in gilded boxes they tend to become golden. The surrounding colour seems to influence the caterpillar "during the twenty hours immediately preceding the last twelve hours of the larval state," "and this is probably the true meaning of the hours during which the caterpillar rests motionless on the surface upon which it will pupate."

"It appears to be certain that it is the skin of the larva which is influenced by surrounding colours during the sensitive period, and it is probable that the effects are wrought through the medium of the nervous system."

Accepting the facts that caterpillars are subtly affected by surrounding colours, so that the quiescent pupæ harmonise with their environment, and that the adjustment has often protective value, we are led to inquire into the origin of this sensitiveness. That the change of colour is not a direct result of external influence is certain, but of the physiological nature of the changes we know little more than that it must be complex. It may be maintained, that "the existing colours and markings are at any rate in part due to the accumulation through heredity of the indirect influence of the environment, working by means of the nervous system;" "to which it may be replied," Poulton continues, "that the whole use and meaning of the power of adjustment depends upon its freedom during the life of the individual; any hereditary bias towards the colours of ancestors would at once destroy the utility of the power, which is essentially an adaptation to the fact that different individuals will probably meet with different environments. As long ago as 1873 Professor Meldola argued that this power of adjustment is adaptive, and to be explained by the operation of natural selection." Poulton's opinion seems to be, that the power of producing variable colouring arose as a constitutional variation apart from the influence of the environment, that the power was fostered in the course of natural selection, and that its limits were in the same way more or less defined in adaptation to the most frequent habitat of the larvæ before and during pupation. The other theory is that the power arose as the result of environmental influence, was accumulated by inheritance throughout generations, and was fostered like other profitable variations by natural selection. The question is whether the power arose in direct relation to environmental influence or not, whether external influence was or was not a primary factor in evolving the power of adapting colour, and in defining it within certain limits.

6. Rapid Change of Colour.—For ages the chamæleon has been famous for its rapid and sometimes striking changes of colour. The members of the Old World genus *Chamæleo* quickly change from green to brown or other tints, but rather in response to physical irritation and varying moods than in relation to change of situation and surrounding colours. So the American "chamæleons" (*Anolis*) change, for instance, from emerald to bronze under the influence of excitement and various kinds of light. Their sensitiveness is exquisite; "a passing cloud may cause the bright emerald to fade." Sometimes they may be thus protected, for "when on the broad green leaves of the palmetto, they are with difficulty perceived, so exactly is the colour of the leaf counterfeited. But their dark shadow is very distinct from beneath." Most of the lizards have more or less of this colour-changing power, which depends on the contraction and expansion of the pigmented living matter of cells which lie in layers in the under-skin, and are controlled by nerves.

In a widely different set of animals—the cuttle-fishes—the power of rapid colour-change is well illustrated. When a cuttle-fish in a tank is provoked, or when one almost stranded on the beach struggles to free itself, or, most beautifully, when a number swim together in strange unison, flushes of colour spread over the body. The sight suggests the blushing of higher animals, in which nervous excitement passing from the centre along the peripheral nerves influences the blood-supply in the skin; but in colour-change the nervous thrills affect the pigment-containing cells or chromatophores, the living matter of which contracts or expands in response to stimulus. It must be allowed that the colour-change of cuttle-fish is oftenest an expression of nervous excitement, but in some cases it helps to conceal the animals.

More interesting to us at present are those cases of colour-change in which animals respond to the hues of their surroundings. This has been observed in some Amphibians, such as tree-frogs; in many fishes, such as plaice, stickleback, minnow, trout, *Gobius ruthensparri*,

Serranus; and in not a few crustaceans. The researches of Brücke, Lister, and Pouchet have thrown much light on the subject. Thus we know that the colour of surroundings affects the animals through the eyes, for blind plaice, trout, and frogs do not change their tint. The nervous thrill passes from eye to brain, and thence extends, not down the main path of impulse—the spinal cord—but down the sympathetic chain. If this be cut, the colour-change does not take place. The sympathetic system is connected with nerves passing from the spinal cord to the skin, and it is along these that the impulse is further transmitted. The result is the contraction or expansion of the pigment in the skin-cells. Though the path by which the nervous influence passes from the eye to the skin is somewhat circuitous, the change is often very rapid. As the resulting resemblance to surroundings is often precise, there can be no doubt that the peculiarity sometimes profits its possessors.

7. **Special Protective Resemblance.** — The likeness between animals and their surroundings is often very precise, and includes form as well as colour. Thus some bright butterflies, *e.g. Kallima*, are conspicuous in flight, but become precisely like brown withered leaves when they settle upon a branch and expose the under sides of their raised wings; the leaf-insects (*Phyllium*) have leaf-like wings and legs; the "walking-sticks" (*Phasmidæ*), with legs thrown out at all angles, resemble irregular twigs; many caterpillars (of *Geometra* moths especially) sit motionless on a branch, supported in a strained attitude by a thin thread of silk, and exactly resemble twigs; others are like bark, moss, or lichen. Among caterpillars protective resemblance is very common, and Mr. Poulton associates its frequent occurrence with the peculiarly defenceless condition of these young animals. "The body is a tube which contains fluid under pressure; a slight wound entails great loss of blood, while a moderate injury must prove fatal." "Hence larvæ are so coloured as to avoid detection or to warn of some unpleasant attribute, the object in both cases being the same—to leave the larva untouched, a touch being practically fatal." Among backboned animals we do not expect to find many

examples of precise resemblance to surrounding objects; but one of the sea-horses (*Phyllopteryx eques*) is said to be exceedingly like the seaweed among which it lives. It is very difficult at present to venture suggestions as to the constitutional tendencies which may have resulted in "walking-leaves" and "walking-sticks," but forms related to these tend to resemble leaves or sticks sufficiently to deter

FIG. 7.—Leaf-insect seated on a branch. (From Belt.)

one from postulating a mere sport as the origin of the peculiarity which distinguishes *Phyllium* or *Phasma*. On the other hand, some of the strangely precise minute resemblances may be the fostered results of slight indefinite sports. It is also possible that some of the cleverer animals, such as spiders, learn to hide among the lichens and on the bark which they most resemble. But in every case, and especially where there are many risks, as among caterpillars, the protective resemblance would be fostered in the course of natural selection.

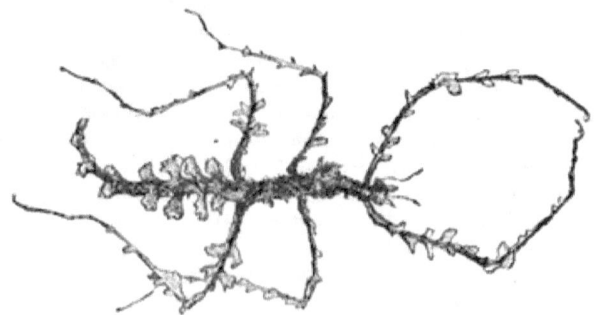

FIG. 8.—Moss insect. (From Belt.)

8. **Warning Colours.**—While many animals are concealed by their colouring, others are made the more conspicuous. But, as the latter are often unpalatable or dangerous, Wallace suggested that the colours were warnings, which, as Poulton says, "assist the education of enemies, enabling them to easily learn and remember the animals which are to be avoided." Expressing the same idea, Belt says, "the skunk goes leisurely along, holding up his white tail as a danger-flag for none to come within range of his nauseous artillery." So, the brightness of the venomous coral-snake (*Elaps*) is a warning; the rattlesnake, excitedly shaking its rattle, "warns an intruder of its presence"; the cobra "endeavours to terrify its enemy by the startling appearance of its expanded hood and conspicuous eye-like marks." The language in which conspicuous colours are described by many naturalists tends to exaggerate the subtlety of animals, for the intentional warning of possible molesters involves rather complex ideas. Belt's description of the skunk, for instance, recalls a more familiar sight—a cat showing fight to a dog—in regard to which Mantegazza gravely tells us that the cat "bristles up her fur, and inflates herself to appear larger, and to frighten the dog who threatens her"! In our desire to be fair to the subtlety of animals, it is indeed difficult to avoid being credulous.

Perhaps the best illustration which Belt gives is that of a certain gaily-coloured frog :—

"In the woods around Santo Domingo there are many frogs. Some are green or brown, and imitate green or dead leaves, and live amongst foliage. Others are dull earth-coloured, and hide in holes and under logs. All these come out only at night to feed, and they are all preyed upon by snakes and birds. In contrast to these obscurely-coloured species, another little frog hops about in the daytime dressed in a bright livery of red and blue. He cannot be mistaken for any other, and his flaming vest and blue stockings show that he does not court concealment. He is very abundant in the damp wood, and I was convinced that he was uneatable so soon as I had made his acquaintance, and saw the happy sense of security with which he hopped about. I took a few specimens home with me, and tried my fowls and ducks with them, but none of them would touch them. At last, by throwing down pieces of meat, for which there was a great competition amongst them, I managed to entice a young duck into snatching up one of the little frogs. Instead of swallowing it, however, it instantly threw it out of its mouth, and went about jerking its head, as if trying to throw off some unpleasant taste."

Admirable, also, are the illustrations given by Mr. Poulton in regard to many caterpillars, such as the larva of the currant or magpie moth (*Abraxas grossulariata*), which is conspicuous with orange and black markings on a cream ground, and is refused altogether, or rejected with disgust, by the hungry enemies of other caterpillars. Professor Herdman and Mr. Garstang have also shown that the Eolid Nudibranchs (naked sea-slugs), with brightly-coloured and stinging dorsal papillæ, are rarely eaten by fishes ; and the same is true of some other conspicuous and unpalatable marine animals.

The general conclusion seems fairly certain that the conspicuousness of many unpalatable or noxious animals is imprinted on the memory of their enemies, who, after paying some premiums to experience, learn to leave animals with "warning colours" alone. It will be interesting to discover how far the bright colour, the nauseous taste, the poisonous properties, the distasteful odour, sometimes found associated, are physiologically related to one another, but to answer these questions we are still unprepared.

9. **Mimicry.**—Mr. Poulton has carefully traced the transition from warning to mimetic appearance, and it is evident that if hungry animals have been so much impressed with the frequent association of unpalatableness and conspicuous colours that they do not molest certain bright and nauseous forms, then there is a chance that palatable forms may also escape if they are sufficiently like those which are passed by. The term mimicry is restricted to those cases

FIG. 9.—Hornet (*Priocnemis*) above, and mimetic bug (*Spiniger*) beneath. (From Belt.)

"in which a group of animals in the same habitat, characterised by a certain type of colour and pattern, are in part specially protected to an eminent degree (the mimicked), and in part entirely without special protection (the mimickers); so that the latter live entirely upon the reputation of the former." The fact was "discovered by Bates in Tropical America (1862), then by Wallace in Tropical Asia and Malaya (1866), and by Trimen in South Africa (1870)"; while Kirby, in 1815, referred to the advantage of a certain fly being like a bee, and of a certain spider resembling an ant.

The constant conditions of mimicry are clearly and tersely summed up by Wallace. They are:—

1. That the imitative species occur in the same area, and occupy the very same station, as the imitated.
2. That the imitators are always the more defenceless.
3. That the imitators are always less numerous in individuals.
4. That the imitators differ from the bulk of their allies.

Fig. 10.—Humming-bird moth (*Macroglossa titan*), and humming-bird (*Lophornis Gouldii*). (From Bates.)

5. That the imitation, however minute, is *external* and *visible* only, never extending to internal characters or to such as do not affect the external appearance.

Many inedible butterflies are mimicked by others quite different. Many longicorn beetles exactly mimic wasps, bees, or ants. The tiger-beetles are mimicked by more harmless insects; the common drone-fly (*Eristalis*) is like a bee; spiders are sometimes ant-like. Mr. Bates relates that he repeatedly shot humming-bird moths in mistake for humming-birds. Among Vertebrates genuine mimicry is rare, but it is well known that some harmless snakes mimic poisonous species. Thus, the very poisonous coral-snakes (*Elaps*), which have very characteristic markings, are mimicked in different localities by several harmless forms. Similarly in regard to birds, Mr. Wallace notices that the powerful "friar-birds" (*Tropidorhynchus*) of Malaya are mimicked by the weak and timid orioles. " In each of the great islands of the Austro-Malayan region there is a distinct species of *Tropidorhynchus*, and there is always along with it an oriole that exactly mimics it."

That there may be mimetic resemblance between distinct forms there can be no doubt, and the value of the resemblance has been verified; but there is sometimes a tendency to weaken the case by citing instances or using terms which have been insufficiently criticised. Thus the facts hardly justify us in saying that the larvæ of the Elephant Hawk Moth (*Chærocampa*) "terrify their enemies by the suggestion of a cobra-like serpent;" or that the cobra, which " inspires alarm by the large eye-like 'spectacles' upon the dilated hood, offers an appropriate model for the swollen anterior end of the caterpillar, with its terrifying markings."

There is only one theory of mimicry, namely, that among the mimicking animals varieties occurred which prospered by being somewhat like the mimicked, and that in the course of natural selection this resemblance was gradually increased until it became dominant and, in many cases, remarkably exact.

As to the primary factors giving rise to the variation, we can only speculate. To begin with, indeed, there must have been a general resemblance between the ancestors of the mimicking animal and those of the mimicked, for cases like the Humming-Bird and its Doppel-Gänger moth are very rare. But this does not take us very far. The beginning of the mimetic change is usually referred to one of those "indefinite," "fortuitous," "spontaneous" variations which are believed to be common among animals. It is logically possible that this may have been the case, and that there was at the very beginning no relation between the variation of the mimicker and the existence of the mimicked. But as illustrations of mimicry accumulate—and they are already

FIG. 11.—Illustrations of Masking. A hermit crab with sea-anemones (after Andres), a crab covered by a sponge (*Suberites*), another crab with seaweed and zoophyte growing on it. (In part after Carus Sterne.)

very numerous—one is tempted to ask whether there may not be in many cases some explanation apart from the action of natural selection upon casual changes. May not the similar surroundings and habits of mimickers and mimicked have sometimes something to do with their resemblance; may it not be that the presence of the mimicked has had a direct, but of course very subtle, influence on the mimickers; is it altogether absurd to suppose that there may be an element of consciousness in the resemblance between oriole and friar-bird?

10. "**Masking**" is one of the most interesting ways in which animals strengthen their hold on life. It is best illustrated on the sea-shore, where there is no little struggle for existence and much opportunity for device. There many animals, such as crabs, are covered by adventitious disguises, so that their real nature is masked. Elsewhere, however, the same may be seen; the cases of the caddis-worms—made of sand particles, small stones, minute shells, or pieces of bark—serve at once for protection and concealment; the cocoons of various caterpillars are often masked by extrinsic fragments. The nests of birds are often well disguised with moss and lichen.

But among marine animals masking is more frequent. "Certain sea-urchins," Mr Poulton says, "cover themselves so completely with pebbles, bits of rock and shell, that one can see nothing but a little heap of stones; and many marine molluscs have the same habits, accumulating sand upon the surface of the shell, or allowing a dense growth of Algæ to cover them."

This masking is in many cases quite involuntary. Thus the freshwater snails (*Lymnæus*) may be so thickly covered with Algæ that they can hardly move, and some marine forms are unable to favour or prevent the growth of other organisms upon their shells. But how far this is from being the whole story is well known to all who are acquainted with our shore crabs. For though they also may be involuntarily masked, there is ample evidence that they sometimes disguise themselves.

The hermit-crabs are to some extent masked within

their stolen shells, especially if these be covered by the Hydroid *Hydractinia* or other organisms. Various other crabs (*Stenorhynchus, Inachus, Maia, Dromia, Pisa*) are masked by the seaweeds, sponges, and zoophytes which cover their carapace. Moreover, the interest of this masking is increased by the fact observed by Mr. Bateson at Plymouth that the crabs sometimes fix the seaweeds for

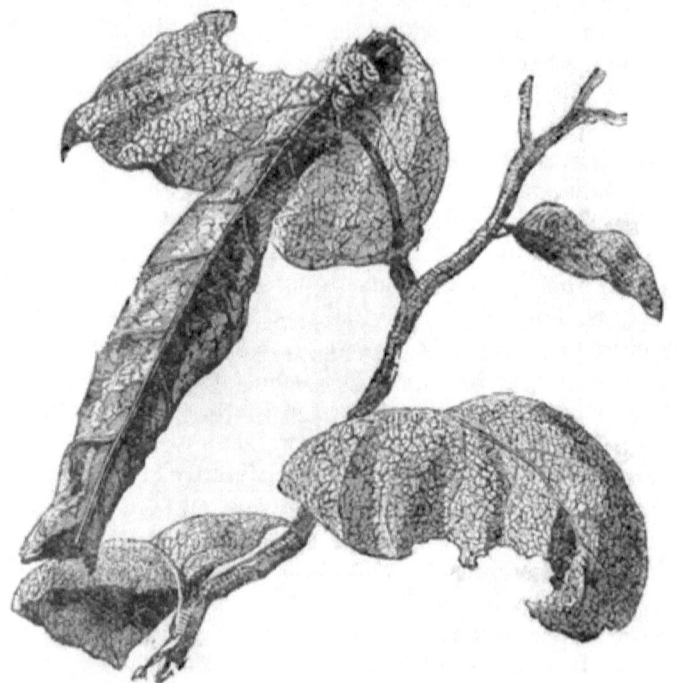

FIG. 12.—Sack-bearing caterpillar (*Saccophora*). (From Bates.)

themselves. Mr. Bateson describes how the crab seizes a piece of weed, tears off a piece, chews the end in his mouth, and then rubs it firmly on his head and legs until it is caught by the curved hairs and fixed. " The whole proceeding is most human and purposeful. Many substances, as hydroids, sponges, Polyzoa, and weeds of many kinds and colours, are thus used; but these various substances are nearly always symmetrically placed on corresponding parts of the body, and particularly long plume-like pieces are fixed on the head." Thus, as Carus Sterne says, is the story of "Birnam's walking wood" re-enacted on the seashore. Furthermore, a *Stenorhynchus* which has been cleaned will immediately begin to clothe itself again, with the same care and precision as before. Mr. Robertson of Millport often saw *Stenorhynchus longirostris*—a common crab—picking about its limbs and conveying the produce to its mouth. "If other observations confirm the view that this animal is a true vegetarian, we shall have one example at least of an independent agriculturist, who is not only superior of his lands, but carries them with him when he removes." I also have seen the crab doing what "the naturalist of Cumbrae" observed. In further illustration of masking we may cite *Dromia vulgaris*, often covered with sponge; *Dromia excavata*, with compound ascidians; the Amphipod *Atylus*, with seaweed; while a species of *Dorippe* is said to bear a bivalve shell, or even a leaf, as a shield, and another crab cuts off the tunic of a sea-squirt and hitches it on his own shoulders.

Sometimes this masking serves as a warning or deterrent; witness that hermit-crab (*Pagurus cuanensis*) whose stolen shell is surrounded by a bright orange sponge (*Suberites domuncula*). As this sponge is full of flinty needles, has a strong odour and a disagreeable taste, we do not wonder that Mr. Garstang finds that fish dislike it intensely, nor can we doubt that the hermit-crab trades on the reputation of its associate. In other cases the masking will aid in concealment and favour attack. To the associations of crabs and sea-anemones we shall afterwards refer.

11. **Combination of Advantageous Qualities.** — Mr. Poulton describes, in illustration of the combination of many methods of defence, the case of the larva of the puss moth (*Cerura vinula*). It resembles the leaves of the poplar and willow on which it lives. When disturbed it assumes a terrifying attitude, mimetic of a Vertebrate appearance! The effect is heightened by the protrusion of two pink whips from the terminal prongs of the body, and finally the creature defends itself by squirting formic acid.

Yet in spite of all this power of defence, the larva often falls a victim to ichneumon-flies. These manage to lay their eggs within the caterpillar, which by and by succumbs to the voracity of the hatched ichneumon maggots. Mr. Poulton believes that the puss moth larva "has been saved from extermination by the repeated acquisition of new defensive measures. But any improvement in the means of defence has been met by the greater ingenuity or boldness of foes; and so it has come about that many of the best-protected larvæ are often those which die in the largest numbers from the attacks of enemies. The exceptional standard of defence has been only reached through the pressure of an exceptional need."

FIG. 13.—"Terrifying attitude" of the caterpillar of *Cerura vinula*. (From Chambers's *Encyclop.*; after Poulton.)

12. **Surrender of Parts.**—Among the strange life-preserving powers which animals exhibit, we must also include that of surrendering parts of the body in the panic of capture or in the struggle to escape. A rat will gnaw off a leg to free itself from a trap, and I have heard of a stoat which did not refrain from amputating more than one limb. But the cases to which we now refer are not deliberate amputations, but reflex and unconscious surrenders. Many lizards (such as our British "slowworm") will readily leave their tails in their captor's grasp; crustaceans, insects, and spiders part with their limbs and scramble off maimed but safe; starfishes, brittle-stars, and feather-stars resign their arms, and the sea-cucumbers their viscera. A large number of cases have been studied by Frédéricq and Giard.

Among Crustacea the habit is most perfectly developed in the crabs, *e.g.* the common shore-crab (*Carcinus mænas*), and in the spiny lobster (*Palinurus*), but it is also exhibited by the crayfish (*Astacus*), the common lobster (*Homarus*), the shrimp (*Crangon*), and the prawn (*Palæmon*). In crabs and in the spiny lobster the surrender of a limb is effected by the forcible contraction of the basal muscles, and the line of rupture is through the second-lowest joint. Frédéricq's researches seem to prove conclusively that the surrender is a reflex and unconscious act, but its protective value is not less great. The chances are in favour of the crab escaping, the residue of muscle prevents hæmorrhage from the stump, and in the course of time the lost limb is replaced by a new growth. The crab does not know what it is doing, but it unconsciously illustrates that it is better that one member should perish than that the whole life should be lost.

Not a few insects readily surrender their legs, but these are not replaced. Spiders are captured if the legs are fixed without irritating the nerves, for that is an essential condition of the reflex amputation. In regard to lizards, also, it has been shown that a reflex nervous excitement, and not mere brittleness, is the condition of surrender. Here, however, the lost tail may be replaced. Among Mollusca a surrender of parts has been recorded of *Harpa ventricosa*, *Doris cruenta*, *Stenopus*, some species of *Helix*, the razor-shell *Solen*; while it is well known that male cuttle-fishes sometimes part with one of their arms for special sexual purposes. A great many "worms" break very easily, and the severed parts are sometimes able to regrow the whole organism.

Among the Echinoderms the tendency to disrupt is exhibited to an extraordinary degree. Thus Professor Preyer has shown that the seven-rayed starfish (*Asterias tenuispina*) surrenders its arms with great readiness, often giving off three or four at a time. But each ray may reproduce an entire starfish. Professor Edward Forbes tells how a specimen of *Luidia*, which he had dredged, was disappearing over the side of the boat when he caught it by one of its arms; it surrendered the arm and escaped, giving "a wink of derision" with one of its eyes. Brittle-stars (Ophiuroids) of many kinds are true to their popular name, and the Crinoids are not less disruptive. Not only are the arms readily given off, but these break into many fragments. There can be no doubt that this habit, combined with the marvellous power of regrowth which these animals possess, is of great protective value, while it is also probable in regard to both Echinoderms and some worms, that the disruption of parts may really increase the number of individuals.

There is no need to enumerate all the protective habits and devices which animals exhibit. Some "feign" death, by falling in panic into a state allied to hypnotic trance, perhaps in some of the higher animals by conscious deception; others roll themselves up into balls, as in forms so different as myriapods and armadillos; but, finally, I shall cite from Dr. Hickson's *Naturalist in North Celebes* one other device. "I often saw advancing slowly over the sea-gardens, in parties of from four to six, a group of cuttle-fish, swimming with an even backward movement, the fringes of their mantles and their arms trembling, and their colour gradually changing to what seemed to me an almost infinite variety of hues as they passed over the various beds of the sea-bottom. Then suddenly there would be a commotion in what was previously a calm and placid scene, the striped and speckled reef fishes would be seen darting away in all directions, and of the cuttle-fishes all that remained were four or five clouds of ink in the clear water. They had thrown dust in the eyes of some small shark or voracious fish."

But I should not like to suggest the idea that animals are always careful and anxious, or forced to continual struggle and shift.

> "They do not sweat and whine about their condition,
> They do not lie awake in the dark and weep for their sins,
> They do not make me sick discussing their duty to God,
> Not one is dissatisfied, not one is demented
> With the mania of owning things;
> Not one kneels to another, nor to his kind that lived thousands of years ago;
> Not one is respectable or unhappy over the whole earth."
> WALT WHITMAN.

CHAPTER V

SOCIAL LIFE OF ANIMALS

1. *Partnerships*—2. *Co-operation and Division of Labour*—3. *Gregarious Life and Combined Action*—4. *Beavers*—5. *Bees*—6. *Ants*—7. *Termites*—8. *Evolution of Social Life*—9. *Advantages of Social Life*—10. *A Note on the Social Organism*—11. *Conclusions*

THE over-fed plant bears many leaves but its flowers are few; the animal which eats too much becomes fat; and we know that within the living body one part may grow out of proportion to the others. It seems as if organ competed with organ within the living engine, as if one tissue outgrew its neighbours in the living web, as if there were some struggle for existence between the individual units which form the city of cells in any of the higher animals. This idea of internal competition has been elaborated by a German biologist, Roux, in a work entitled *The Struggle of Parts within the Organism*, and it is full of suggestiveness. It can be verified from our own experience; but yet it seems strange. For we rightly think of an organism as a unity in which the parts are bound together in mutual helpfulness, being members one of another.

Now, just as a biologist would exaggerate greatly if he maintained that the struggle of parts was the most important fact about an organism, so would a naturalist if he maintained that there was in nature struggle only and no helpfulness.

Coherence and harmony and mutual helpfulness of parts—whether these be organs, tissues, or cells—are certainly facts in the life of individuals; we have now to see how far the same is true of the larger life in which the many are considered as one.

1. **Partnerships.**—Animals often live together in strange partnerships. The "beef-eater" birds (*Buphagus*) perch on cattle and extract grubs from the skin; a kind of plover (*Pluvianus ægyptius*) removes leeches and other parasites from the back of the crocodile, and perhaps "picks his teeth," as Herodotus alleged; the shark is attended by the pilot-fish (*Naucrates ductor*), who is shielded by the shark's reputation, and seems to remove parasites from his skin.

Especially among marine animals, we find many almost constant associations, the meaning of which is often obscure. Two gasteropods *Rhizochilus* and *Magilus* grow along with certain corals, some barnacles are common on whales, some sponges and polypes are always found together, without there being in any of these cases either parasitism or partnership. But when we find a little fish living contentedly inside a large sea-anemone, or the little pea-crab (*Pinnotheres*) within the horse-mussel, the probable explanation is that the fish and the crab are sheltered by their hosts and share their food. They are not known to do harm, while they derive much benefit. They illustrate one kind of "commensalism," or of eating at the same table.

But the association between crabs and sea-anemones affords a better illustration. One of the hermit-crabs of our coast (*Pagurus prideauxii*) has its borrowed shell always enveloped by a sea-anemone (*Adamsia palliata*), and *Pagurus bernhardus* may be similarly ensheathed by *Adamsia rondeletii*. Möbius describes two crabs from Mauritius which bear a sea-anemone on each claw, and in some other crabs a similar association occurs. It seems that in some cases the crab deliberately chooses its ally and plants it on its shell, and that it does not leave it behind at the period of shell-changing. Deprived of its polype companion, one was seen to be restlessly ill at ease until it obtained another of the same kind. The use of the sea-anemone as a mask to the crab—and also perhaps as aid in attack or defence—is obvious; on the other hand, the sea-anemone is carried about by the crab and may derive food from the crumbs of its bearer's repast.

Commensalism must be distinguished from parasitism, in which the one organism feeds upon its host, though it is quite possible that a commensal might degenerate into a parasite. Quite distinct also is that intimate partnership known as symbiosis, illustrated by the union of algoid and fungoid elements to form a lichen, or by the occurrence of minute Algæ as constant internal associates and helpful partners of Radiolarians and some Cœlenterates.

2. **Co-operation and Division of Labour.**—The idea of division of labour has been for a long time familiar to men, but its biological importance was first satisfactorily recognised by Milne-Edwards in 1827.

Among the Stinging-animals there are many animal colonies, aggregates of individuals, with a common life. These begin from a single individual and are formed by prolific budding, as a hive is formed by the prolific egg-laying of a queen-bee. The mode of reproduction is asexual in the one case, sexual in the other; the resulting individuals are physically united in the one case, psychically united in the other; but these differences are not so great as they may at first sight appear. Many masses of coral are animal colonies, but among the members or "persons," as they are technically called, division of labour is very rare; moreover, in the growth of coral the younger individuals often smother the older. In colonial zoophytes the arborescent mode of growth usually obviates crushing; and there is sometimes very marked division of labour. Thus in the colony of *Hydractinia* polypes, which is often found growing on the shells tenanted by hermit-crabs, there may be a hundred or more individuals all in organic connection. The polypes are minute tubular animals, connected at their bases, and stretching out from the surface of the shell into the still water of the pool in which the hermit-crab is resting. But among the hundred individuals there are three or four castes, the differences between which probably result from the fact that in such a large colony perfect uniformity of nutritive and other conditions is impossible. Individuals which are fundamentally and originally like one another grow to be different, and perform different functions according to the caste to which they belong.

Many are nutritive in form like the little freshwater *Hydra*—tubular animals with an extensile body and with a terminal mouth wreathed round by mobile tentacles. On these the whole nutrition of the colony depends. Beside these there are reproductive "persons," which cannot feed, being mouthless, but secure the continuance of the species and give rise to embryos which start new colonies. Then there are long, lank, sensitive members, also mouthless, which serve as the sense-organs of the colony, and are of use in detecting food or danger. When danger threatens, the polypes cower down, and there are left projecting small hard spines, which some regard as a fourth class of individuals—starved, abortive members like the thorns on the hawthorn hedge. In recognising their utility to the colony as a whole we can hardly overlook the fact that their life as individuals is practically nil. They well illustrate the dark side of division of labour.

FIG. 14.—Colony of *Hydractinia echinata*. *a*, nutritive individuals; *b*, reproductive individuals; *c*, abortive spines; and there are also long mouthless individuals specialised in sensitiveness. (From Chambers's *Encyclop.*; after Allman).

Herbert Spencer and Ernst Haeckel have explained very clearly one law of progress among those animals which form colonies. The crude form of a colony is an *aggregate* of similar individuals, the perfected colony is an *integrate* in which by division of labour greater harmony of life has resulted, and in which the whole colony is more thoroughly compacted into a unity. Among the Stinging-animals, we find some precise illustrations of such integrated colonies, especially in the Siphonophora of which the Portuguese Man-of-War (*Physalia*) is a good example. There is no doubt that these beautiful organisms are colonies of individuals, which in structure are all referable to a "medusoid" or jellyfish-like type. But the division of labour is so harmonious, and the compacting or organisation of the colony is so thorough, that the whole moves and lives as a single organism.

E. Perrier in his work entitled *Les Colonies Animales* (Paris, 1882), shows how organic association may lead from one grade of organisation and individuality to another, and explains very clearly how sedentary and passive life tends to develop mere aggregates, while free and active life tends to integrate the colony. With this may be compared A. Lang's interesting study on the influence of sedentary life and its connection with asexual reproduction—*Das Einfluss des Festsitzen* (Jena, 1889). Haeckel, in his *Generelle Morphologie* (2 vols., Berlin, 1866), was one of the first to shed a strong clear light on the difficult subject of organic individuality, its grades and its progressive complexity. To Spencer, *Principles of Biology* (2 vols., London, 1863-67), we owe in this connection the elucidation of the transition from aggregates to integrates, and of the lines of differentiation, *i.e.* the progressive complication of structure which is associated with division of labour.

3. **Gregarious Life and Combined Action.**—Most mammals are in some degree gregarious. The solitary kinds are in a distinct minority. The isolated are exposed to attack, the associated are saved by the wisdom of their wisest members and by that strength which union gives. Many hoofed animals, such as deer, antelopes, goats, and elephants, live in herds, which are not mere crowds, but organised bands, with definite conventions and with a power of combined resistance which often enables them to withstand the attacks of carnivores. Marmots and prairie-dogs, whose "cities" may cover vast areas, live peaceful and prosperous lives. Monkeys furnish many illustrations of successful gregarious life. As individuals most of them are comparatively defenceless, and usually avoid coming to close quarters with their adversaries; yet in a body they are formidable, and often help one another out of scrapes. Brehm tells how he encountered a troop of baboons which defied his dogs and retreated in good order up the

heights. A young one about six months old being left behind called loudly for aid. "One of the largest males, a true hero, came down again from the mountain, slowly went to the young one, coaxed him, and triumphantly led him away—the dogs being too much astonished to make an attack."

FIG. 15.—Chimpanzee (*Anthropopithecus* or *Troglodytes calvus*). (From Du Chaillu.)

Many birds, such as rooks and swallows, nest together, and the sociality is often advantageous. Kropotkine cites from Dr. Coues an observation in regard to some little cliff-swallows which nested in a colony quite near the home of a prairie-falcon. "The little peaceful birds had no fear of their rapacious neighbour; they did not let it even approach to their colony. They immediately surrounded it and chased it, so that it had to make off at once." Of the cranes, Kropotkine notes that they are extremely "sociable and live in friendly relations, not only with their congeners, but also with most aquatic birds." They post sentries, send scouts, have many friends and few enemies, and are very intelligent. So is it also with parrots. " The members of each band remain faithfully attached to each other, and they share in common good or bad luck." They feed together, fly together, rest together ; they send scouts and post sentinels ; they find protection and pleasure in combination. Like the cranes, they are very intelligent, and safe from most enemies except man.

On the other hand, some of the most successful carnivores, *e.g.* wolves, hunt in packs, and not a few birds of prey (some eagles, kites, vultures) unite to destroy their quarry. Combination for defence has its counterpart in combination for attack. In both cases the collective action is often associated with the custom of posting sentinels, who warn the rest, or of sending scouts to reconnoitre. Peculiarly interesting are those cases in which the relatively weak unite to attack the strong ; thus a few kites will rob an eagle, and wagtails will persecute a sparrow-hawk. Kropotkine has noticed how the aquatic birds which crowd on the shores of lakes and seas often combine to drive off intruding birds of prey. " In the face of an exuberant life, the ideally armed robber has to be satisfied with the off-fall of that life."

Among many animals there is co-operation in labour, as well as combination for attack or defence. Brehm relates that baboons and other monkeys act in thorough concert in plundering expeditions, sending scouts, posting sentinels, and even forming a long chain for the transport of the spoil. It is said that several Hamadryad baboons will unite to turn over a large stone, sharing the booty found underneath. When the Brazilian kite has seized a prey too large for it to carry, it summons its friends; and Kropotkine cites a remarkable case in which an eagle called others to the carcase. Pelicans fish together in great companies, forming a wide half-circle facing the shore and catching the fish thus enclosed. Burial beetles unite to bury the dead mouse or bird in which the eggs are laid, and the dung-beetles help one another in rolling balls of food. But of all cases of combined activity the migration of birds is at once the most familiar and the most beautiful—the gathering together, the excitement before starting, the trial flights, the reliance placed in the leaders. Migration is usually social, and is sustained by tradition.

4. **Beavers.**—That the highly-socialised beavers have been exterminated in many countries where they once abounded is no argument against their sociality, for man has ingenuity enough to baffle any organisation. A family of about six members inhabits one house, and in suitable localities—secluded and rich in trees—many families congregate in a village community. The young leave the parental roof in the summer of their third year, find mates for themselves, and establish new homesteads. The community becomes overcrowded, however, and migrations take place up and down stream, the old lodges being sometimes left to the young couples. It is said, moreover, that lazy or otherwise objectionable members may be expelled from the society, and condemned to live alone. Under constraint of fear or human interference, and away from social impulse, beavers may relapse into lazy and careless habits, and in many cases each family lives its life apart; but in propitious conditions their achievements are marvellous. The burrow may rise into a constructed home, and the members of many families may combine in wood-cutting and log-rolling, and yet more markedly in constructing dams and digging canals. Make allowances for the exaggeration of enthusiastic observers, but read Mr. Lewis Morgan's stories of the evolution of a broken burrow into a comfortable lodge, varying according to the local conditions; of the adaptation of the dams against the rush of floods; of canals hundreds of feet in length—labours without reward until they are finished; of the short-cut waterways across loops of the river; and of "locks" where continuous canals are, from the nature of the ground, impossible. The Indians have invested beavers with immortality, but it is enough for us to recognise that

they exhibit more sagacity than can be explained by hereditary habit, for they often adapt their actions to novel conditions in a manner which must be described as intelligent. Especially when we remember that the beaver belongs to a somewhat stupid rodent race, are we inclined to believe that it is the cleverest of its kind because the most socialised.

5. **Bees.**—Many centuries have passed since men first listened to the humming of the honey-bees, and found in the hive a symbol of the strength of unity. From Aristotle's time till now naturalists have been studying the life of bees, without exhausting either its facts or its suggestions. The society is very large and complex, yet very stable and successful. Its customs seem now like those of children at play, and now like the realised dreams of social reformers. The whole life gives one the impression of an old-established business in which all contingencies have been so often experienced that they have ceased to cause hesitation or friction. There is indeed much mortality, some apparent cruelty, and the constantly recurring adventure of migration; but though hive may war against hive, inter-civic competition has virtually ceased, and the life proceeds smoothly with the harmony and effectiveness of a perfected organisation.

G. 16.—Honey-bee (*Apis mellifica*). A, queen; B, drone; C, worker. (From Chambers's *Encyclop.*)

The mother-bee, whom we call a "queen"—though she is without the wits and energy of a ruler—is to this extent head of the community, that, by her prolific egg-laying, she increases or restores the population. Very sluggish in their ordinary life are the numerous males or "drones," one of whom, fleet and vigorous beyond his fellows, will pair with a queen in her nuptial flight, himself to die soon after, saved at least from the expulsion and massacre which await all the sex when the supplies of honey run short in autumn. The queen and drones are important only so far as multiplication is concerned. The sustained life of the hive is wholly in the hands of the workers, who in brains, in activity, and general equipment are greatly superior to their "queen." "The queen has lost her domestic arts, which the worker possesses in a perfection never attained by the ancestral types; while the worker has lost her maternal functions, although she still possesses the needed organs in a rudimentary state."

What a busy life is theirs, gathering nectar and pollen unweariyingly, while the sunshine lasts, neatly slipping into the secrets of the flowers or stealing their treasures by force, carrying their booty home in swift sweeping flight, often over long distances unerringly, unloading the pollen from their hind-legs and packing it into some cells of the comb, emptying out the nectar from their crop or honey-sac into store-cells, and then off again for more—such is their socialised mania for getting. But, besides these "foragers"—for the most part seniors—there are younger stay-at-home "nurses," whose labours, if less energetic, are not less essential. For it is their part to look after the grubs in their cradles, to feed them at first with a "pap" of digested nectar, and then to wean them to a diet of honey, pollen, and water; to attend the queen, guiding her movements and feeding her while she lays many eggs, sometimes 2000 to 3000 eggs in a day. Mr. Cheshire, in his incomparably careful book on *Bees and Beekeeping*, laughs at the "many writers who have given the echo to a mediæval fancy by stating that the queen is ever surrounded by a circle of dutiful subjects, reverently watching her movements, and liable to instant banishment upon any neglect of duty. These it was once the fashion to compare to the twelve Apostles, and, to make the ridiculous suggestion complete, their number was said to be invariably twelve!" But Mr. Cheshire's own account of the nurses' work, and of the whole life of the hive, is more marvellous than any mediæval fancy.

We have not outlined nearly all the labours of the workers. There is the exhausting though passive labour of forming the wax which oozes out on the under-surface of the body, and then there is the marvellous comb-building, at which the bees are very neat and clever workers, though they do not deserve the reputation for mathematical insight once granted them. "Their combs," Mr. Cheshire says, "are rows of rooms unsurpassably suitable for feeding and nurturing the larvæ, for giving safety and seclusion during the mystic sleep of pupa-hood, for ensconcing the weary worker seeking rest, and for safely warehousing the provisions ever needed by the numerous family and by all during the winter's siege. Corridors run between, giving sufficient space for the more extensive quarters of the prospective mother, and affording every facility to the busy throng walking on the ladders the edges of their apartments supply ; while the exactions of modern hygiene are fully met by air, in its native purity, sweeping past the doorway of every inhabitant of the insect city."

We shall not seek to penetrate into the more hidden mysteries of the life of bees ; for instance, " how the drones have a mother but no father," or how high feeding makes the difference between a queen and a worker. An outline of the yearly life is more appropriate. From the winter's rest the surviving bees reawaken when the early-flowering trees begin to blossom ; the workers engage in a "spring cleaning," and the queen restores the reduced population by egg-laying. New supplies of food are brought in, new bees are born, and in early summer we see the busy life in all its energy. The pressure of increased population makes itself felt, and migration or "swarming" becomes imperative. In due time and in fair weather "the old mother departs with the superabundance of the population." Meanwhile in the parent-hive drones have been born, and several possible queens await liberation. The first to be set free has to hold her own against newcomers, or it may be to die before one of them. The successful new queen soon becomes restless, issues forth in swift nuptial flight, is fertilised by a drone, and returns to her home to begin prolific egg-laying, and perhaps after a time to lead off another swarm. During the busy summer, when food is abundant, the lazy males are tolerated; but when their function is fulfilled, and when the supplies become scarce, they are ruthlessly put to death. "No sooner does income fall below expenditure, than their nursing sisters turn their executioners, usually by dragging them from the hive, biting at the insertion of the wing. The drones, strong for their especial work, are, after all, as tender as they are defenceless, and but little exposure and abstinence is required to terminate their being. So thorough is the war of extermination, that no age is spared; even drone eggs are devoured, the larvæ have their juices sucked and their 'remains' carried out—a fate in which the chrysalids are made to take part, the maxim for the moment being, He that will not work, neither shall he eat." This Lycurgan tragedy over, the equilibrium of the hive is more secure, and the winter comes.

The social life of hive-bees is of peculiar interest, because it represents the climax of a series of stages. Hermann Müller has traced the plausible history of the honey-bee from an insect like the sand-wasp, and has shown in other kinds of bees the various steps by which the pollen-gathering and nectar-collecting organs have been developed. The habits of life gradually lead up to the consummately social life of the hive. Thus *Prosopis*, which lays its eggs in the pith of bramble-stems; the wood-boring *Xylophaga*; and the leaf-cutting *Megachile*, which lines its burrows with circles cut from rose leaves, are *solitary* bees. The various species of humble- or bumble-bee (*Bombus*), so familiarly industrious from the spring, when the willows bear their catkins, till the autumn chill benumbs, are halfway to the hive-bees; for they live in societies of mother, drones, and workers during summer, while the sole surviving queens hibernate in solitude. From the humble-bee, moreover, we gain this hint, that the home is centred in the cradle, for it is in a nest with honey and pollen stored around the eggs that the hive seems to have begun.

6. **Ants.**—Even more suggestive of our own social organ-

isation is the Liliputian world of the ants, who, like microscopic men, build barns and lay up stores, divide their labour and indulge in play, wage wars and make slaves. Like the bee-hive, the ant-nest includes three kinds of individuals— a queen mother or more than one, a number of short-lived males, and a crowd of workers. The queen is again pre-eminently maternal, and, if we can trust the enthusiastic observers, she is attended with loyal devotion, not without some judicious control. Farren White describes how the workers attend the queen in her perambulations: "They formed round her when she rested; some showed their regard for her by gently walking over her, others by patiently watching by her and cherishing her with their antennæ, and in every way endeavouring to testify to their affectionate attachment and generous submission." Gould ventures further, alleging that "in whatever apartment a queen condescends to be present, she commands obedience and respect, and a universal gladness spreads itself through the whole cell, which is expressed by particular acts of joy and exultation. They have a peculiar way of skipping, leaping, and standing up on their hind legs, and prancing with the others. These frolics they make use of both to congratulate each other when they meet, and to show their regard for the queen." These are wonderful lists of assumed emotions! Should an indispensable queen be desirous to quit the nest, the workers do not hesitate, it is said, to keep her by force, and to tear off her wings to secure her stay. It is certain at least that as the queens settle down to the labour of maternity, their wings are lost—perhaps in obedience to some physiological necessity. From the much greater number of the wingless workers, we are apt to forget that the males and mothers of the social ants are winged insects; but this fact becomes impressive if in fine summer weather we are fortunate enough to see the males and young queens leaving the nest in the nuptial flight, during which fertilisation takes place. Rising in the air they glitter like sparks, pale into curling smoke, and vanish. " Sometimes the swarms of a whole district have been noticed to unite their countless myriads, and, seen at a distance, produce an effect resembling the flashing of the Aurora Borealis; sometimes the effect is that of rainbow hues in the spray of laughing waterfalls; sometimes that of fire; sometimes that of a smoke-wreath." "Each column looks like a kind of slender network, and has a tremulous undulating motion. The noise emitted by myriads and myriads of these creatures does not exceed the hum of a single wasp. The slightest zephyr disperses them." After this midsummer day's delight of love, death awaits many, and sometimes most. The males are at best short-lived, but the surviving queens, settling down, may begin

FIG. 17.—Saüba ants at work; to the left below, an ordinary worker; to the right a large-headed worker; above, a subterranean worker. (From Bates.)

to form nests, gathering a troop of workers, or sometimes proceeding alone to found a colony.

A caste of workers (*i.e.* normally non-reproductive females) distinct from the males and queens, involves, of course, some division of labour; but there is more than this. Workers of different ages perform different tasks— foraging or housekeeping, fighting or nursing, as the case may be; and just as the various human occupations leave marks both for good and ill in those who follow them, so the division of labour among ants is associated with differences of structure. Thus, in the Saüba or Umbrella Ant of Brazil (*Œcodoma cephalotes*), so well described by Bates in his *Naturalist on the Amazons*, there are three classes of workers. All the destructive labour of cutting sixpence-like disks from the leaves of trees is done by individuals with small heads, while others with enormously large heads simply walk about looking on. These "worker-majors" are not soldiers, nor is there any need for supervising officers. "I think," Bates says, "they serve, in some sort, as passive instruments of protection to the real workers. Their enormously large, hard, and indestructible heads may be of use in protecting them against the attacks of insectivorous animals. They would be, on this view, a kind of *pièces de résistance*, serving as a foil against onslaughts made on the main body of workers." The third order of workers includes very strange fellows, with the same kind of head as the worker-majors have, but "the front is clothed with hairs instead of being polished, and they have in the middle of the forehead a twin simple eye," which none of the others possess. Among the honey ants (*Myrmecocystus mexicanus*) described by Dr. M'Cook from the "Garden of the Gods" in Colorado, the division of labour is almost like a joke. The workers gather "honey" from certain galls, and discharge their spoils into the mouths of some of their stay-at-home fellows. These passive "honey-pots" store it up, till the abdomen becomes tense and round like a grape, but eventually they have even more tantalisingly to disgorge it for other members of the community. But this habit of feeding others is exhibited, as Forel has shown, by many species of ants. The hungry apply to the full for food, and get it. A refusal is said to be sometimes punished by death !

Marvellous in peace, the ants may also practise the anti-social "art of war," sometimes against other communities of the same species, sometimes with other kinds. "Their battles," Kirby says, "have long been celebrated; and the date of them, as if it were an event of the first importance, has been formally recorded." Æneas Sylvius, after giving a very circumstantial account of one contested with great obstinacy between a great and small species on the trunk of a pear tree, gravely states, "This action was fought in the pontificate of Eugenius IV., in the presence of Nicholas Pistoriensis, an eminent lawyer, who related the whole history of the battle with the greatest fidelity." In the fray the combatants are thoroughly absorbed, yet at a little distance other workers are uninterruptedly treading their daily paths; the mêlée is intense, yet every ant seems to know those of its own party; the result of it all is often nothing. We laugh at the ants—the laugh comes back on ourselves.

In some cases an expedition has the definite end of slave-making, as is known to be true of *Formica sanguinea*—a British species, and of *Polyergus rufescens*, found on the Continent. The former captures the larvæ of *Formica fusca*, carries them home, and owns them henceforth as well-treated slaves; while the Amazon Ant (*Polyergus*) draws its supply from both *F. fusca* and *F. cunicularia*, and seems to have become almost dependent on its captives. Indeed, Hüber says that he never knew the Amazons take nourishment but from the mouth of the negro captives; while Lubbock notes that every transition exists between bold and active baron-like marauders and enervated masters, who are virtually helpless parasites upon their slaves—a suggestive illustration of laziness outwitting itself.

Slaves somewhat painfully suggest domesticated animals, and these are also to be found among ants. For what Linnæus said long ago, that the ants went up trees to "milk their cows, the Aphides," is true. The ants tickle these little plant-lice with their antennæ, and lick the juice which oozes from them; nay more, according to some, they inclose and tend these milch kine, and even breed them at home. Seed-harvesting and the like may be fairly called agricultural, and do not the leaf-cutters grow mushrooms, or at least feed on the fungi which grow on the leaves, stored some say with that end in view? The driver ants, "whose dread is upon every living thing," when they are on the stampede, remind us of the ancient troops of nomad hunters, though some of them are blind. Thus there are hunting, agricultural, and pastoral ants—three types, as Lubbock remarks, offering a strange analogy to the three great phases in the history of human development.

Very quaint is another habit of this "little people, so exceeding wise,"—that of keeping or tolerating guests in the home! These are mostly little beetles, and have been carefully studied by Dr. Weismann, who distinguishes true guests (*Atemeles, Lomechusa, Claviger*) which are cared for and fed by the ants, from others (*Dinarda, Hæterius, Formicoxenus*) which are tolerated, though not treated with special friendliness, and which feed on dead ants or vegetable débris; while a third set are tolerated—like mice in our houses—only because they cannot be readily turned out. Of the genuine guests, the best known is *Atemeles*, a lively animal, constantly moving its feelers, and experimenting with everything. If one be attacked by a hostile ant, it first seeks to pacify its antagonist by antennary caresses, but if this is hopeless it emits a strong odour, which seems to narcotise the ant. These little familiars are really dependent upon their hosts, who feed them and get caresses in return. It is easy to understand the presence of pests in the ants' home, but *Atemeles* and *Lomechusa* are pets, taken away by the owners when there is a flitting, and exhibiting, as Lubbock also observes, "international relations," since they can be shifted from one nest to another, or even from species to species. It seems likely enough, as Emery suggests, that these semi-domesticated pets are moralised intruders, and, like our cats, they seem to retain some of their original traits.

I cannot linger longer over the interesting characteristics of ants, though I should like to speak of their architecture, of their roads, tunnels, bridges, and covered ways; of their care for the young, and sometimes even for the disabled; of their proverbial industry, and yet of their indulgence in "sportive exercise." It would be profitable to think about the contrast between solitary ants (*Mutilidæ*) who have no "workers," and the complex life of a community in which there are half a million residents; or about their æsthetic sensitiveness, for they see light and hear sound for which our eyes and ears are not adapted; or about their power of recognising their fellow-citizens (even when intoxicated), and of communicating definite impressions to one another by a subtle language of touch and gesture; or about their instincts and intelligence, and the limitations of these. But it will be better to read some of the detailed observations, endeavouring, though necessarily with slight success, to think into the nature of ants,—their pertinacity, their indomitable "pluck," their tireless industry, their organic sociality. Surely all will agree with Sir John Lubbock, to whose patient observations we owe so much, that, "when we see an ant-hill, tenanted by thousands of industrious inhabitants, excavating chambers, forming tunnels, making roads, guarding their home, gathering food, feeding the young, tending their domestic animals, each one fulfilling its duties industriously and without confusion, it is difficult altogether to deny them the gift of reason," or, perhaps more accurately, intelligence, for we cannot escape the conviction " that their mental powers differ from those of men not so much in kind as in degree."

Kropotkine says that the work of ants is performed "according to the principles of voluntary mutual aid." " Mutual aid within the community, self-devotion grown into a habit, and very often self-sacrifice for the common welfare, are the rule." The marvels of their history are " the natural outcome of the mutual aid which they practise at every stage of their busy and laborious lives." To this mode of life is also due " the immense development of individual initiative." Ants are not well protected, but "their force is in mutual support and mutual confidence." "And if the ant stands at the very top of the whole class of Insects for its intellectual capacities; if its courage is only equalled by the most courageous Vertebrates, and if its brain—to use Darwin's words—'is one of the most marvellous atoms of matter in the world, perhaps more so than the brain of man,' is it not due to the fact that mutual aid has entirely taken the place of mutual struggle in the communities of ants?"

7. **Termites.**—The true ants are so supremely interest-

ing, that the Termites or "white ants" (which are not ants at all) are apt to receive scant justice. Perhaps inferior in intelligence, they have the precedence of greater antiquity and all the interest which attaches to an old-established society. Nor is their importance less either to practical men or to speculative biologists. In 1781 Smeathman gave some account of their economy, noting that there were in every species three castes; "first, the working insects, which, for brevity, I shall generally call *labourers*; next, the fighting ones or *soldiers*, which do no kind of labour; and, last of all, the winged ones, or *perfect insects*, which are male and female, and capable of propagation."

The "workers," blind and wingless, and smallest in the ant-hill, do all the work of foraging and mining, attending the royal pair and nursing the young. The soldiers, also blind and wingless, are much larger than the workers, but there are relatively only a few in each hill. "They stand," Prof. Drummond says, "or promenade about as sentries, at the mouths of the tunnels. When danger threatens, in the shape of true ants, the soldier termite advances to the fight." "With a few sweeps of its scythe-like jaws it clears the ground, and while the attacking party is carrying off its dead, the builders, unconscious of the fray, quietly continue their work." At home, in the ant-hill, shut up in a chamber whose door admits workers but is much too small for the tenants to pass out if they would, a fortunate investigator sometimes finds the royal pair. The male is sometimes even larger than the soldier, and is in many ways different, though by no means extraordinary. The queen-mother, however, is a very strange organism. She measures two to six inches, while the worker is only about a fifth of an inch in length. Like her mate, she sees, and she once had wings like his, but they have dropped off. The hind part of the body is enormously distended with eggs, and "the head bears about the same proportion to the rest of the body as does the tuft on his Glengarry bonnet to a six-foot Highlander." In her passivity and "phenomenal corpulence," she is a sort of *reductio ad absurdum* of femaleness —"a large, cylindrical package, in shape like a sausage, and as white as a bolster." But have some admiration for her: she sometimes lays 60 eggs per minute, or 80,000 in a day, and continues reproducing for months. As she lays, she is assiduously fed by the nursing-workers, while the eggs are carried off to be hatched in the nurseries. At the breeding season, numerous winged males and females leave

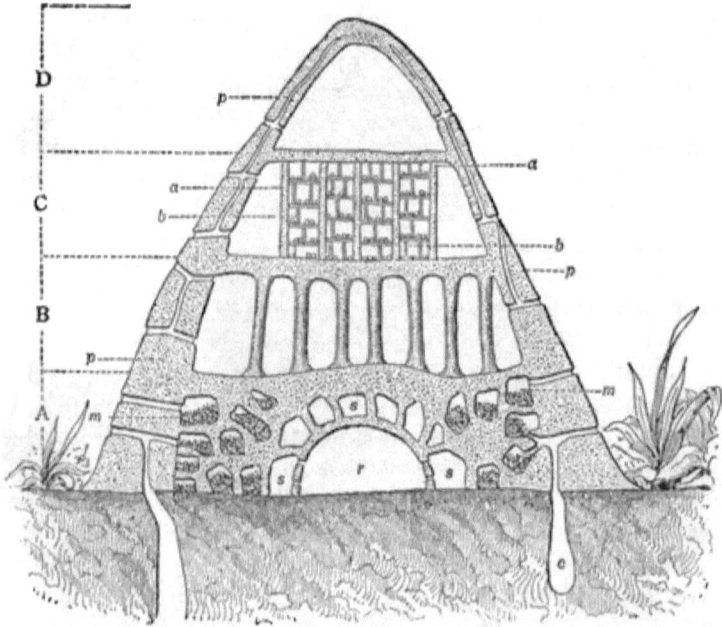

FIG. 18.—Diagrammatic section of a termite's nest (after Houssay). In the walls there are winding passages (*p*); uppermost is a well-aired empty attic (D) the next story (C) is a nursery where the young termites are hatched on shelves (*a*) and (*b*); the next is a hall (B) supported by pillars; beneath this is a royal chamber (*r*) in which the king and queen are imprisoned; around this the chambers of worker-termites (*s*) and some store-chambers (*m*); excavated in the ground are holes (*c*) out of which the material used in making the termitary was dug. The whole structure is sometimes 10-15 feet in height.

the hill and its workers in swarms, most of them simply to die, others to mate with individuals from another hill and to begin to form new colonies.

The plot of the story becomes more intricate, however, when we notice Fritz Müller's observations, that " besides the winged males and females which are produced in vast numbers, and which, leaving the termitary in large swarms, may intercross with those produced in other communities, there are (in some if not all of the species) wingless males and females which never leave the termitary where they are born, and which replace the winged males or females whenever a community does not find, in due time, a true king or queen.". There is no doubt as to the existence of both winged and wingless royal pairs. According to Grassi, the former fly away in spring, the others ascend the throne in summer. The complementary kings or viceroys die before winter ; their mates live on, widowed but still maternal, till at least the next summer.

This replacement of royalty reminds us that hive-bees, bereft of their queen, will rear one from the indifferent grub, but the termites with which we are best acquainted seem almost always to have a reserve of reproductive members. This other difference between termites and ants or bees should be noticed, that in the latter the "workers" are highly-developed, though sterile females, while in the former the workers seem to be arrested forms of both sexes. They are children which do not grow up.

8. **Evolution of Social Life.**—To Professor Alfred Espinas both naturalists and sociologists are greatly indebted for his careful discussion of the social life of animals. It may be useful, therefore, to give an outline of the mode of treatment followed in his work—*Des Sociétés Animales : Étude de Psychologie Comparée* (Paris, 1877) :—

Co-operation, which is an essential characteristic of all society, implies some degree of organic affinity. There are, indeed, occasional associations between unrelated forms—" mutualism," in which both associates are benefited ; "commensalism," in which the benefit is mainly one-sided ; parasitism, which is distinctly anti-social, deteriorating the host and also the rank of the temporarily benefited parasite. Of normal societies whose members are mutually dependent, two kinds may be distinguished—(*a*) the organically connected colonies of animals, in which there is a common nutritive life ; (*b*) those associations which owe their origin and meaning to reproduction. Of the latter, some do not become more than domestic, and these are distinguished as conjugal (in which the parents alone are concerned), maternal (in which the mother is the head of the family), and paternal (in which the male becomes prominent). But higher than the pair and the family is what Espinas calls the "peuplade," what we usually call the society, whose bonds are, for the most part, psychical.

But let us consider this problem of the evolution of sociality. The body of every animal—whether sponge or mammal—is a city of living units or cells. But there are far simpler animals than sponges. The very simplest animals, which we call firstlings or Protozoa, differ from all the rest, in being themselves units. The simplest animals are single cells; each is comparable to one of the myriad units which make up a sponge, a coral, a worm, a bird, a man.

Here, therefore, there is an apparent gulf. The simplest animals are units—single cells; all other animals are combinations of units—cities of cells. How is this gulf to be bridged? It is strange that evolutionists have not thought more about this, for on the transition from a unit to a combination of units the possibility of higher life depends.

Every higher animal begins its individual life as a single cell, comparable to one of the firstlings. This single cell, or egg-cell, divides; so do most of the Protozoa. But when a Protozoon divides, the results separate and live independent lives; when an egg-cell divides, the results of division cohere. Therefore, the whole life of higher animals depends upon a coherence of units.

But how did this begin? What of the gulf between single-celled Protozoa and all the other animals which are many-celled? Fortunately we are not left to mere speculation. The gulf has been bridged, else we should not exist; but, more than that, the bridge, or part of it, is still left. There are a few of the simplest animals which form loose colonies of units, which, when they divide, remain together. Whether it was through weakness, as I am inclined to believe, that the transition forms between Protozoa and higher animals became strong, or for some hidden reason, we do not know. Some speak of this coherence of firstlings as a primal illustration of organic association, co-operation, surrender of individuality, of sociality at a low level, but it is unwise to apply these words to creatures so simple. All that we certainly know is that some of the simplest animals form loose colonies of units, that the gulf between them and the higher animals is thus bridged, and that the bridging depends on coherence. Our first con-

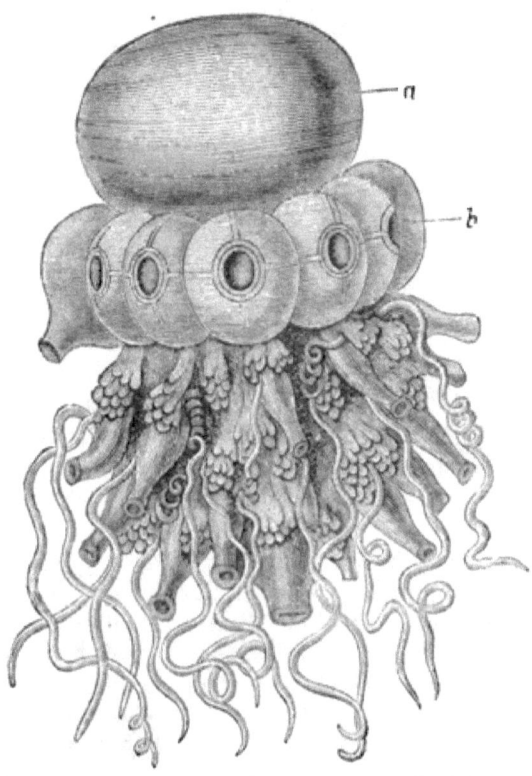

FIG. 19.—Siphonophore colony, showing the float (*a*), the swimming-bells (*b*); the nutritive, reproductive, and other members of the colony beneath. (From the *Evolution of Sex*; after Haeckel.)

clusion, therefore, is, that the possibility of there being any higher animals depends, primarily at least, not on competition but on the coherence of units.

Our next step is this: When we study sponges, or zoophytes, or most corals, or some types usually classed as "worms," we see that the habit of forming colonies is common. Every sponge is a simple sac to begin with, but it buds off others like itself, and the result is a coherent colony. A zoophyte is not one individual, but a connected colony of individuals. Throughout the colony there is one life; all the individuals have a common origin, and all are members one of another. In varying degrees of perfection the life of the whole is unified. Moreover, the unity is often increased, not diminished, by the fact that the individuals are not all alike. There is division of labour among them; some may feed while others reproduce, some feel much while others may be quite callous. Thus, as we already mentioned, the Portuguese Man-of-War, a colony of small jellyfish-like individuals, has much division of labour, and yet there is much, though by no means perfect, unity of life.

Our second conclusion is that among many animals—beginning with sponges and ending with the sea-squirts, which are acknowledged to be animals of high degree—the habit of forming colonies is common, and that these colonies, though organically continuous, illustrate the essence of society; for in them many individuals of common descent and nature are united in mutual dependence and helpfulness.

The next step towards an understanding of the social relations of animals is very different from that in which we have recognised the habit of forming colonies. The factor which we have now to acknowledge is the love of mates. This also has its history, this also has its prophecies among the firstlings, but we shall simply assume as a fact that among crustaceans and insects first, in fishes and amphibians afterwards, in reptiles too, but most conspicuously among birds and mammals, the males are attracted to the females, and in varying degrees of perfection enter into relations of mutual helpfulness. The relations and the attractions may be crude enough to begin with, but perhaps even we hardly know to what heights of devotion their highest expressions may attain. To mere physical fondness are added subtler attractions of sight and hearing, and these are sublimed in birds and mammals to what we call love. This love of mates broadens out; it laps the family in its folds; it diffuses itself as a saturating influence through the societies of animals and of men. "Sociability," Espinas says, "is based on the friendliness of mates."

The fourth step is the evolution of the family. From monkeys and beavers and many kinds of birds, to ants and bees and diverse insects, many animals illustrate family life. There is no longer the physical continuity characteristic of the colony, but there is a growing psychical unity. It is natural that the first ties of family life should be those between mother and young, and should be strongest when the number of offspring is not very large. But even in some beetles, and more notably in certain fishes and amphibians, the males exhibit parental care and affection; while in higher animals, especially among birds, the parents often divide the labours of the family. "Children," Lucretius said, "children with their caresses broke down the haughty temper of parents."

The fifth step is the combination of families into a society, such as we find illustrated by monkeys and beavers, cranes and parrots, and in great perfection by ants. The members are less nearly related than in the family, but there may be even more unity of spirit.

I do not say that it is easy to understand how coherence of units led to the formation of a "body," how colonies became integrated and the labours of life more and more distributed, how love was evolved from apparently crude attractions between the sexes, how the love of mates was broadened into parental and filial affection, or how families well knit together formed the sure foundations of society; but I believe that it is useful to recognise these steps in the history.

We hardly know how to express ourselves in regard to the origin of affection. But I cannot get beyond Aristotle's fundamental principle of evolution, that there is nothing in the end which was not also in the beginning.

Yet we may fairly say that the sociality and helpfulness of animals are flowers whose roots are in kinship. Offspring are continuous in nature with their parents; the family has a unity though its members be discontinuous and scattered; "the race is one and the individual many."

9. Advantages of Social Life.—But animals are social, not only because they love one another, but also because sociality is justified of her children. "The world is the abode of the strong," but it is also the home of the loving; "contention is the vital force," but the struggle is modified and ennobled by sociality.

(a) *Darwin's Position.*— Darwin observed that "the individuals which took the greatest pleasure in society would best escape various dangers; while those that cared least for their comrades, and lived solitary, would perish in greater numbers." He distinctly emphasised that the phrase "the struggle for existence" was to be used in a wide and metaphorical sense—to include all the endeavours which animals make both selfishly and unselfishly to strengthen their foothold and that of their offspring. But he was not always successful in retaining this broad view, nor was he led to compute with sufficient care to what extent mutual aid is a factor in evolution counteractive of individualistic struggle.

Without losing sight of the reality of the struggle for existence; without disputing the importance of natural selection as a condition of evolution—securing that the relatively fittest changes succeed; without ignoring what seems almost a truism, that love and social sympathies have also been fostered in the course of natural selection; we maintain—(1) that many of the greatest steps of progress—such as those involved in the existence of many-celled animals, loving mates, family life, mammalian motherhood, and societies—were not made by the natural selection of indefinite variations; (2) that affection, co-operation, mutual helpfulness, sociality, have modified the struggle for material subsistence by lessening its intensity and by ennobling its character.

(b) *Kropotkine's Position.*—Against Prof. Huxley's conclusion that "Life was a continual free-fight, and beyond the limited and temporary relations of the family the

Hobbesian war of each against all was the normal state of existence," let me place that of Kropotkine, to whose admirable discussion of mutual aid among animals I again acknowledge my indebtedness.

" Life in societies is no exception in the animal world. It is the rule, the law of nature, and it reaches its fullest development with the higher Vertebrates. Those species which live solitary, or in small families only, are relatively few, and their numbers are limited. . . . Life in societies enables the feeblest mammals to resist, or to protect themselves from, the most terrible birds and beasts of prey; it permits longevity; it enables the species to rear its progeny with the least waste of energy, and to maintain its numbers, albeit with a very slow birth-rate; it enables the gregarious animals to migrate in search of new abodes. Therefore, while fully admitting that force, swiftness, protective colours, cunning, and endurance of hunger and cold, which are mentioned by Darwin and Wallace as so many qualities making the individual or the species the fittest under certain circumstances, we maintain that under *any* circumstances sociability is the greatest advantage in the struggle for life. . . . The fittest are thus the most sociable animals, and sociability appears as the chief factor of evolution, both directly, by securing the well-being of the species while diminishing the waste of energy, and indirectly by favouring the growth of intelligence. . . . Therefore combine—practise mutual aid! That is the surest means for giving to each and to all the greatest safety, the best guarantee of existence and progress—bodily, intellectual, and moral. That is what nature teaches us."

10. **A Note on " The Social Organism."**—It is common nowadays to speak of society as " the social organism," and the metaphor is not only suggestive but convenient —suggestive because it is profitable to biologist and sociologist alike to follow out the analogies between an organism and society, convenient because there is among organisms —in aggregates like sponges, in perfected integrates like birds—a variety sufficient to meet all grades and views of society, and because biologists differ almost as much in

their conceptions of an "organism" as sociologists do in regard to "society."

It may be questioned, however, whether we need any other designation for society than the word society supplies, and whether the biological metaphor, with physical associations still clinging to it, is not more illusory than helpful. For the true analogy is not between society and an individual organism, but between human society and those incipient societies which were before man was. Human society is, or ought to be, an integrate—a spiritual integrate—of organisms, of which the bee-hive and the ants' nest, the community of beavers and the company of monkeys, are like far-off prophecies. And in these, as in our own societies, the modern conception of heredity leads us to recognise that there is a very real unity even between members physically discontinuous.

The peculiarity of human society, as distinguished from animal societies, depends mainly on the fact that man is a social person, and knows himself as such. Man is the realisation of antecedent societies, and it is man's realisation of himself as a social person which makes human society what it is, and gives us a promise of what it will be. As biologists, and perhaps as philosophers, we are led to conclude that man is determined by that whole of which he is a part, and yet that his life is social freedom; that society is the means of his development, and at the same time its end; that man has to some extent realised himself in society, and that society has been to some extent realised in man.

But I am slow to suppose that we, who in our ignorance and lack of coherence are like the humbler cells of a great body, have any adequate conception of the social organism of which we form part.

11. **Conclusions.**— I would in the main agree with Kropotkine that "sociability is as much a law of nature as mutual struggle"; with Espinas that " Le milieu social est la condition nécessaire de la conservation et du renouvellement de la vie"; and with Rousseau that "man did not make society, but society made man."

CHAPTER VI

THE DOMESTIC LIFE OF ANIMALS

1. *The Love of Mates*—2. *Love and Care for Offspring*

WINTER in our northern climate sets a spell upon life. The migrant birds escape from it, but most living things have to remain spell-bound, some hiding with the supreme patience of animals, others slumbering peacefully, others in a state of "latent life" stranger than death. But within the hard rind of the trees, or lapped round by bud scales, or imprisoned within the husks of buried seeds, the life of plants is ready to spring forth when the south wind blows; beneath the snow lie the caterpillars of summer butterflies, the frogs are waiting in the mud of the pond, the hedgehog curled up sleeps soundly, and everywhere, under the seeming death, life rests until the spring. "For the coming of Ormuzd, the Light and Life Bringer, the leaf slept folded, the butterfly was hidden, the germ concealed, while the sun swept upwards towards Aries."

But when spring does come, heralded by returning migrants—swallows and cuckoos among the rest—how marvellous is the reawakening! The buds swell and burst, the corn sends up its light green shoots, the primrose and celandine are in blossom, the mother humble-bee comes out from her hiding-place and booms towards the willow catkins, the frogs croak and pair, none the worse of their fast, the rooks caw noisily, and the cooing of the dove is heard from the wood. Then, as the pale flowers are succeeded by those of brighter tints, as the snowy hawthorn gives place to the laburnum's "dropping wells of fire" and the bloom of the lilac, the butterflies flit in the sunshine, the chorus of birds grows stronger, and the lambs bleat in the valley. Temperature rises, colours brighten, life becomes strong and lusty, and the earth is filled with love.

1. **The Love of Mates.**—In human life one of the most complex musical chords is the love of mates, in the higher forms of which we distinguish three notes—physical, emotional, and intellectual attraction. The love of animals, however, we can only roughly gauge by analogy; our knowledge is not sure enough to appreciate it justly, though we know beyond any doubt that in many the physical fondness of one sex for another is sublimed by the addition of subtler emotional sympathies. Among mammals, which frequently pair in spring, the males are often transformed by passion, the "timid" hare becomes an excited combatant with his rivals, while in the beasts of prey love often proves itself stronger than hunger. There is much ferocity in mammalian courtship—savage jealousy of rivals, mortal struggles between them, and success in wooing to the strongest. In many cases the love-making is like a storm—violent but passing. The animals pair and separate—the females to motherhood, the males to their ordinary life. A few, like some small antelopes, seem to remain as mates from year to year; many monkeys are said to be monogamous; but this is not the way of the majority.

Birds are more emotional than mammals, and their love-making is more refined. The males are almost always more decorative than their mates, and excel in the power of song. They may sing, it is true, from sheer gladness of heart, from a genuine joy of life, and their lay rises "like the sap in the bough"; but the main motive of their music is certainly love. It may not always be music to us, but it is sweet to the ears for which it is meant—to which in many tones the song says ever "Hither, my love! Here I am! Here!" Nor do the male birds woo by singing alone, but by love dances and by fluttering displays

CHAP. VI *The Domestic Life of Animals*

II

of their bright plumage; with flowers, bright pods, and shining shells, the bower-birds decorate tents of love for their honeymoon. The mammals woo chiefly by force; the birds are often moved to love by beauty, and mates often live in prolonged partnership with mutual delight and helpfulness. Sixty years before Darwin elaborated his theory of sexual selection, according to which males have grown more attractive because the most captivating suitors were most successful in love, the ornithologist Bechstein noted how the female canary or finch would choose the best singer among a crowd of suitors; and there seems some reason to believe that the female's choice of the most musical or the most handsome has been a factor in progress. Wallace, on the contrary, maintains that the females are plainly dressed because of the fate which has befallen the conspicuous during incubation, and surely they must thus be handicapped. To others it seems more natural to admit that there is truth in both Darwin's and Wallace's conclusions, but to regard the males as stronger, handsomer, or more musical simply because they are males, of more active constitutional habit than their mates. To this view Mr. Wallace himself inclines.

Compared with the lion's thunder, the elephant's trumpeting, or the stag's resonant bass, and the might which lies behind these, or with the warble of the nightingale, the carol of the thrush, the lark's blithe lay, or the mocking-bird's nocturne, and the emotional wealth which these express, the challenges and calls of love among other classes of animals are apt to seem lacking in force or beauty. But our human judgment affords no sure criterion. The frogs and newts, which lead on an average a somewhat sluggish life, wake up at pairing time, and croak according to their strength. The males are often furnished with two resonating sacs at the back of the mouth, and how they can croak dwellers by marsh-land know; the North American bullfrog bellows by himself, and the South American tree-frogs hold a concert in the branches.

Of the mating of fishes we know little, but there are some well-known cases alike of display and of tournament. The stickleback fights with his rivals, leads his mate to the nest by captivating wiles, dances round her in a frenzy,

FIG. 21.—Male and female bird of paradise (*Paradisea minor*). (From *Evolution of Sex*; after Catalogue of Dresden Museum.)

and afterwards guards the eggs with jealous care. The

male salmon, with their hooked lower jaws, fight with their rivals, sometimes to the death.

Among insects the love-play is again very lively. Like birds, many of these active animals are very beautiful in colour and form, especially in the male sex, and a display of charms has often been noticed. Like birds, though in a different fashion, some of them are musical, using their hard legs and wing-edges as instruments. The crickets chirp merrily, the cicadas "sing," and the death-watch taps at the door of his mate.

In the summer night, when colours are put out by the darkness, the glow-worm shines brightly on the mossy bank. In the British species (*Lampyris noctiluca*) the winged male and the wingless female are both luminous; the latter indeed excels in brightness, while her mate has larger eyes. Whatever the phosphorescence may mean to the constitution of the insect, it is certainly a love-signal between the sexes. But we know most about the Italian glow-worm (*Luciola italica*), of whose behaviour we have a lively picture—thanks to Professor Emery's nocturnal observations in the meadows around Bologna. The females sit among the grass; the males fly about in search of them. When a female catches sight of the flashes of an approaching male, she allows her splendour to shine. He sees the female's signal, and is swiftly beside her, circling round like a dancing elf. But one suitor is not enough. The female attracts a levée. In polite rivalry her devotees form a circle and await the coquette's choice. In the two sexes, Emery says, the colour of the light is identical, and the intensity seems much the same, though the love-light of the female is more restricted. The most noteworthy difference is that the luminous rhythm of the male is more rapid, with briefer flashes; while that of the female is more prolonged, with longer intervals, and more tremulous—illumined symbols of the contrast between the sexes.

While recognising the genuinely beautiful love-making of most birds, we did not ignore that the courtship of most mammals is somewhat rough. So, after admiring the love dances of many butterflies, the merry songs of the grasshoppers, and the flashing signals of the glow-insects, it is just that we should turn to the strange courtship of spiders, which is less ideal. Of what we may be prepared to find we get a hint from a common experience. Not long ago I found in a gorge some spiders which I had never seen before. Wishing to examine them at leisure, I captured a male and a female, and, having only one box, put them, with misgivings, together. When I came to examine them, however, the male was represented by shreds. Such unnatural conduct, though by no means universal among spiders, is common. The tender mercies of spiders are cruel. We have lately obtained an account of the courtship of spiders from George W. and Elizabeth G. Peckham, from whose careful observations I select the following illustrations:

According to these observers, "there is no evidence that the male spiders possess greater vital activity; on the contrary, it is the female that is the more active and pugnacious of the two. There is no relation in either sex between development of colour and activity. The *Lycosidæ*, which are the most active of all spiders, have the least colour-development, while the sedentary orb-weavers show the most brilliant hues. In the numerous cases where the male differs from the female by brighter colours and ornamental appendages, these adornments are not only so placed as to be in full view of the female during courtship, but the attitudes and antics of the male spider at that time are actually such as to display them to the fullest extent possible. The fact that in the *Attidæ* the males vie with each other in making an elaborate display, not only of their grace and agility, but also of their beauty, before the females, and that the females, after attentively watching the dances and tournaments which have been executed for their gratification, select for their mates the males that they find most pleasing, points strongly to the conclusion that the great differences in colour and in ornament between these spiders are the result of sexual selection."

These conclusions support Darwin's position that the female's choice is a great factor in evolving attractiveness, and are against Wallace's contention that bright colours express greater vitality, and that the females are less brilliant because enemies eliminate the conspicuous. It is quite likely that Darwin's view is true in some cases (*e.g.* these spiders), and Wallace's conclusion true in others (*e.g.* birds and butterflies), or that both may be true in many cases; while the fact that the males of these spiders are always more brilliant than their mates suggests again that the brilliancy is wrapped up along with the mystery of maleness, which it is not sufficient to define merely as superabundant vitality, or as greater activity, but rather as a tendency towards a relative increase of destructive or disruptive vital changes over those which are constructive or conservative. But the problem is very complex, and dogmatic conclusions are premature. We need to know the chemical nature and history of the pigments to which the colour is due; we need to have an approximate balance-sheet of the income and expenditure of the two sexes. Enough of this, however; let us return to the pictures. We talk about romance— listen to these patient observers:

FIG. 22.—Two male spiders (*Habrocestum splendens* to the left, and *Astia vittata* to the right) displaying themselves before their mates. (After G. W. and E. G. Peckham.)

"On reaching the country we found that the males of *Saitis pulex* were mature and were waiting for the females, as is the way with both spiders and insects. In this species there is but little difference between the sexes. On May 24th we found a mature female and placed her in one of the larger boxes, and the next day we put a male in with her. He saw her as she stood perfectly still, twelve inches away. The glance seemed to excite him, and he at once moved toward her. When some four inches from her he stood still, and then began the most remarkable performances that an amorous male could offer to an admiring female. She eyed him eagerly, changing her position from time to time, so that he might always be in view. He, raising his whole body on one side by straightening out the legs, and lowering it on the other by folding the first two pairs of legs up and under, leaned so far over as to be in danger of losing his balance, which he only maintained by sidling rapidly toward the lowered side. The palpus, too, on this side was turned back to correspond to the direction of the legs nearest it. He moved in a semicircle of about two inches, and then instantly reversed the position of the legs and circled in the opposite direction, gradually approaching nearer and nearer to the female. Now she dashes toward him, while he, raising his first pair of legs, extends them upward and forward as if to hold her off, but withal slowly retreats. Again and again he circles from side to side, she gazing toward him in a softer mood, evidently admiring the grace of his antics. This is repeated until we have counted one hundred and eleven circles made by the ardent little male. Now he approaches nearer and nearer, and when almost within reach whirls madly around and around her, she joining and whirling with him in a giddy maze. Again he falls back, and resumes his semicircular motions with his body tilted over; she, all excitement, lowers her head and raises her body, so that it is almost vertical. Both draw nearer, she moves slowly under him, he crawling over her head, and the mating is accomplished." The males are quarrelsome and fight with one another; but after watching "hundreds of seemingly terrible battles" between the males of twelve different species, the observers were forced to the conclusion that "they are all sham affairs gotten up for the purpose of displaying before the females, who commonly stand by interested spectators."

"It seemed cruel sport at first to put eight or ten males (of *Dendryphantes capitatus*) into a box to see them fight,

FIG. 23.—Two male spiders (*Zygoballus bettini*) fighting. (After G. W. and E. G. Peckham.)

but it was soon apparent that they were very prudent little fellows, and were fully conscious that 'he who fights and runs away will live to fight another day.' In fact, after two weeks of hard fighting we were unable to discover one wounded warrior. . . . The single female (of *Phidippus morsitans*) that we caught during the

Fig. 24.—Male argus pheasant displaying its plumage. (From Darwin.)

summer was a savage monster. The two males that we provided for her had offered her only the merest civilities when she leaped upon them and killed them." "The female of *Dendryphantes elegans* is much larger than the male, and her loveliness is accompanied by an extreme irritability of temper, which the male seems to regard as a constant menace to his safety; but his eagerness being great, and his manners devoted and tender, he gradually overcomes her opposition. Her change of mood is only brought about after much patient courting on his part." In other species (*Philæus militaris*) the males take possession of young females and keep guard over them until they become mature. We sometimes hear of courtship by telephone. In the Epeiridæ spiders "it seems to be carried on, to some extent at least, by a vibration of web lines," as M'Cook and Termeyer have also observed.

Surely it is a long gamut this, from a mammal's clamant call and forcible wooing, or from the sweet persuasiveness of our singing birds, and the fluttering displays of others, to the trembling of a thread in the web of a spider. But, however varied be the pitch of the song and the form of the dance, all are expressions of love.

Mates are also attracted to one another by odours. These are best known in mammals (*e.g.* beaver and civet) and in reptiles; they predominate in the males, and at the breeding season. They usually proceed from skin glands; but we understand little about them. They serve as incense or as stimulant, but perhaps this usefulness is secondary. The zoologist Jaeger regards the odoriferous substances in plants and animals as characteristic of and essentially associated with each life; but without going so far we may recognise that in the general life of flowers and animals alike odours are very important. We know, too, that certain odours make much impression upon us; such as those of hawthorn and of the hay-field, of newly-mown grass and of withered leaves, of violet and of lavender; and furthermore, that in some mysterious way some fragrances excite or soothe the system, and have become associated with sexual and other emotions.

2. **Love and Care for Offspring.** — Gradual as the incoming of spring has been the blossoming of parental love among animals. We cannot tell in what forms it first appeared in distinctness. We cannot say Lo here! or Lo there! for it is latent in them all.

In many of the lower animals the units which begin new lives are readily separated from the parent; but in others, *e.g.* some of the simplest, or some by no means simple "worms," and even some insects, the parent life disappears in giving birth to the young. Reproduction or the continuance of the species often involves a sacrifice of the individual life.

It is strangely true, even in the highest forms, that reproduction, though a blossoming of the whole life, is also the beginning of death. It is costly, and brings death as well as life in its train. This is tragically illustrated by many insects, such as butterflies, who die soon after reproducing, though often not before they have, in obedience to instinctive impulse, cared most effectively for their eggs—the results of which they do not live to see. Think also of the mayflies, or Ephemeridæ, who, after a prolonged aquatic life as larvæ, become winged, dance in the sunlight for an hour, mate and reproduce, and die.

Picture the long larval life in the water, and the short aerial happiness lasting for an evening or two. Long life, compared with the span of many other insects, but short love; there may be years of patience, and but a day of pleasure; great preparations, and the anti-climax of death. The eggs lie half conscious in the water, faintly stirred by the growing life within, lapped round about by peace,—though the trout thin them sorely. In the survivors the embryos become conscious, awaken from their rocking, and turn themselves in their cradles. See the larvæ creep forth, wash themselves gaily in the water, and hungrily fall upon their prey, some smaller insects. The little "water-wings" grow, and the air soaks into the blood; the larvæ cast their skins many times, and hide from the fishes. At length comes the final moult, and the making of the air-wings, of which in the summer evening you may see the first short flight as the insects rise like a living mist from the pool. But even yet a thin veil, too truly suggestive of a shroud, encumbers them; and they rest wearily on the grass or on the branches of the willow. Watch them writhe and jerk, as if impatient, till at length their last encumbrance — their "ghost," as naturalists call it — is thrown off. Now the other life, the life of love, begins. Merrily they dance up and down, dimpling the smooth water into smiling with a touch — chasing, embracing, separating. See the filmy fairy wings, the large lustrous eyes of the males, the tail filaments gracefully sweeping in the dance! They never pause to eat — they could not if they tried; **hunger is past, love is present, and in the near future is death.** The evening shadows grow longer, — shadows of death to the Ephemerides. The trout jump at them, a few rain-drops thin the throng, the stream bears others away. The mothers lay their eggs in the water, and wearily die forthwith — cradle and tomb are side by side; the males seem to pass in a sigh from the climax of loving to the other crisis of dying. But the eggs are in the water, and the dance of love is more than a dance of death. Turning homewards, we cannot but think sadly of other Ephemerides, of patient larval life, of the gradual revealing of the higher self, of shrouds thrown aside and wedding robes put on, of hunger eaten up by love, of the sacrifice of maternity, of cradle and tomb together. Yet we remember the eggs in the water, the promise of the future beneath the surface of the stream. Under the horse-chestnut tree, too, the wind has blown the shed petals like white foam, but the tree itself is strong like Ygdrasil, and among the branches a bird sings in the twilight.

Returning in more matter-of-fact mood to parental care, we need not dwell upon those cases where the young are simply sheltered for a while about the body of the mother, hanging to a jellyfish, on some sea-urchins hidden in tents of spines, in one or two sea-cucumbers half buried in the skin, adhering to the naked ventral surface of the common little leech (*Clepsine*), imprisoned in modified tentacles in some marine worms, carried about in a dorsal brood-chamber in many water-fleas, or under the curved tail of higher crustaceans, retained within the gills of bivalves, and so on. Such adaptations are interesting, they involve prolonged physical contact between mother and offspring, but we are in search of cases where the parent acts as if she cared for her young.

But this care, as we said, begins very gradually. Thus, in some lowly crustaceans the young may return to the brood-chamber of the mother, even after hatching and moulting; and young crayfish are said to return to the shelter of the maternal tail after they have been set adrift. Strange, too, are the *males* of some sea-spiders (Pycnogonida), who carry about the ova on their legs. It is confidently stated that the headless freshwater mussel keeps the embryos imprisoned even after the normal period, until some freshwater fish be present, to which they may attach themselves; while some cuttle-fishes are said to exert themselves in keeping their egg-clusters clean and safe.

But it is among insects, with their full, free life, that we see the best examples of parental care in backboneless animals. Some scoff at the "beetle-pricker" or the scarabeist,—and such genial laughter as that of the *Professor at the Breakfast Table* has a healthy resonance,—but those who scoff have not read Kirby's *Letters*, else they would feel that the student of insects watches at a well-head of romance and marvel inexhaustibly fresh. What, for instance, shall we say of the worker-bees, who, though no parents, tend and nurse the grubs with constant care; or of the likewise sexless worker-ants, whose first endeavour when the nest is disturbed is to save, not themselves, but the young; or of the care that flies, moths, and other insects will take to lay their eggs in substances and situations best fitted for the future young? We must think back into the past history of climatic and other conditions if we would understand the frequently elaborate provision which mother insects make for offspring which they never see; the ancestors had probably a longer life, and had the gratification of seeing the result of their labours, and now the inherited habit works on, perhaps with no vision of the future. We must also allow that the offspring mistakenly deposited by an imperfect maternal instinct would most likely die, and thus leave the race more select. But after thinking out these explanations, the facts remain marvellous. Thus W. Marshall saw an Ichneumon fly (*Polynema natans*) remain twelve hours under water, without special adaptations for such a life, swimming about with her wings, and depositing her eggs within the larvæ of caddis-flies!

We are accustomed, the same naturalist says, to look upon a hen which gathers her brood under her wings as a picture of loving care, but we must recognise that the same is true of earwigs, spiders, and scorpions. Many of us have lifted a large stone on the dry bank, and seen the hurry-scurry of small animals; there are earwigs among the rest, and the pale-yellowish young crowd quickly under the shelter of their mothers, who stand guard with open pincers. Female spiders, too, so fierce and impatient as mates, are most "respectable mothers." Some make nests, guard, feed, and even fight for the young; others carry the eggs about with them. "I have often," Marshall says, "made fun of the little creatures, taking away their precious egg-sac and removing it to a slight distance. It was interesting to see how eagerly they sought, and how joyously, one may even say, they sprang upon their 'one and all' when they found it again. Sometimes I cheated them with a little ball of wool of the size, form, and colour of the egg-sac, which they quickly seized, and as rapidly rejected."

Many fishes lay their eggs by hundreds in the water, and thenceforth have nothing more to do with them, but even among these cold-blooded animals there are illustrations of parental care. From a bridge over the river you may be able to watch the female salmon ploughing a furrow in the gravelly bed, and there laying her eggs, careful not to disturb the places where others have already spawned. In quiet by-pools you may find the gay male stickleback guarding the nest which he has made of twined fibres partly glued together with mucus. There the female has laid eggs, but he has driven her forth: he will do all the nursing himself. No approaching enemy is too large for him to attack; his courage equals his seeming pride. When the young are hatched, but not yet able to fend for themselves, his cares are increased tenfold. It is hard to keep the youngsters in the cradle. "No sooner has he brought one bold truant back, than two others are out, and so it goes on the whole day long."

We are not clever enough to understand why the males among many fishes are so much more careful than the females. For the stickleback is not alone in his excellent behaviour. The male Chinese macropod (*Polyacanthus*) makes a frothy nest of air and mucus, in which he places his mate's eggs. He, too, watches jealously over the brood, and "has his hands—or rather his mouth—full to recover the hasty throng when they stray, and to pack them again into their cradle." Of all strange habits, perhaps that is strangest which some male fish (*e.g. Arius*) have of hatching the eggs in their mouths; what external dangers must have threatened them before this quaint brooding-chamber was chosen! Or is it not almost like a joke to see the male sea-horse swelling up as the eggs which he has stowed away in an external pocket hatch and mature, "till one day we see emerging from the aperture a number of small, almost transparent creatures, something like marks of interrogation." But some female fishes also carry their eggs about, attached to the ventral surface (in the Siluroid fish, *Aspredo*), or stowed away in a ventral pouch (in *Solenostoma*, allied to pipe-fishes), arrangements which recur among amphibians, but on the dorsal surface of the body.

FIG. 25.—Sea-horse (*Hippocampus guttulatus*). (From *Evolution of Sex*; after Atlas of Naples Station.)

Amphibians, like fishes, to which they are linked by many ties, are either quaint or careless parents. Again, the males assume the responsibilities of nurture. The obstetric frog (*Alytes obstetricans*), common in some parts of the Continent, takes the eggs from his mate, winds them round his hind-legs, and retires into a hole, whence, after a fortnight or so, he betakes himself to the water, there to be relieved by the speedy hatching of his precious burden. Even quainter is the habit of the male of a Chilian frog (*Rhinoderma darwinii*), who keeps the eggs and the young in a pouch near the larynx, turning a resonating sac in a most matter-of-fact way into a cradle. He is somewhat leaner after it is all over. It is interesting to notice how similar forms and habits recur among animals of different kinds, like the theme in some musical compositions. The spiral form of shell common in the simple chalk-forming Foraminifers recurs in the pearly nautilus; the eye of a fish is practically like that of many a cuttle, though the two are made in quite different ways; and an extraordinary development of paternal care may signalise animals so distinct as sea-spider, stickleback, and frog.

But we must not be unfair to the female amphibians. Without doubt most of them are willing to be quickly rid of their eggs or young, and as these are usually very numerous, the mortality in the pools is of little moment. In some cases, however, water-pools are less available than in Britain, and then we find adaptations securing the welfare of the young. The black salamander of the Alps, living at elevations where pools are rare, retains her twin offspring until more than half of the tadpole life is past. They breathe and feed in a marvellous way within the body of the mother, and are born as lung-breathers. In the case of the Surinam Toad (*Pipa*), the male places half a hundred eggs on the back of the female, where they become surrounded by small pockets of skin, from which the young toads writhe out fully formed. In two other cases (*Nototrema* and *Notodelphys*), the above somewhat expensive adaptation, which involves a great destruction of skin, is replaced by a dorsal pouch in which the eggs hatch, an arrangement dimly suggestive of the pouch of kangaroos and other marsupial mammals.

Fishes and amphibians are linked closely by their likeness in structure, and, as we have seen, they are somewhat alike in parental habits; but how great is the contrast between the habits of birds and reptiles, in spite of their genuine blood-relationship. Yet the python coiled round her eggs is a prophecy of the brooding birds, as in past ages the flopping Saurians prophesied their swift-winged flight. The sharpness of the contrast is also lessened by the fact that a few birds, like the mound-builders, do not brood at all; while others, it must be confessed, are somewhat careless. But, exceptions and criminals apart, birds are so lavish in their love, so constant in their carefulness, that it is difficult to speak of them without exaggeration. I am quite willing to allow that they often act without thought (that is half the beauty of it); nor do I doubt that many species would have gone to the wall long since in the struggle of life if the parents had not taken so much care of the young; but I would rather emphasise at present the reality that they do sacrifice themselves for the sake of their young to a most remarkable degree, and spend themselves not for individual ends, but for their offspring.

Before the time of egg-laying the birds build their nests, eagerly but without hurry, instinctively yet with some plasticity, and often with much beauty. On the laid eggs, which require warmth to develop, the mothers brood, and though to rest after reproduction is natural, the brooding is not without its literal patience. Among polygamous birds the males are, as one would expect, more or less careless of their mates, but most of the monogamous males are careful either in sharing the duty of brooding or in supplying the females with food. After the eggs hatch, the degree of care required varies according to the state of the young; for many are precociously energetic and able to look after themselves, while others still require prolonged nurture. They need large quantities of food, to supply which all the energies of both parents seem sometimes no more than adequate; they may still require to be brooded over, and certainly to be protected from rain and enemies. After they are reared, they have to be taught to fly, to catch food, to avoid danger, and a dozen other arts. With what apparent love—willing and joyous —is all this done for them!

CHAP. VI *The Domestic Life of Animals* 113

Consider the cunning often displayed in leaving or approaching the nest, in removing débris which would betray the whereabouts of the young, or in distracting attention to a safe distance; remember, too, that some birds

FIG. 26.—Nest of tailor-bird (*Orthotomus benettii*). (After Brehm.)

will shift either eggs or young to a new resting-place when extreme danger threatens; estimate the energy spent in feeding the brood, sometimes on a diet quite different from that of adult life; and acknowledge that the parental instinct is very deeply rooted, since fostering young not their own may be practised by orphaned birds of both sexes. Listen to the bird which has been bereaved, and tell me is not the "lone singer wonderful, causing tears"?

The female of the Indian and African hornbill nests in a hole in a tree, the entrance to which she plasters up so that no room is left either for exit or entrance. The

I

Malays imagined that this was the work of the jealous male, but it is the female's own doing. "She sits," Marshall says, "securely hidden, safe from any carnivore or mischievous ape or snake stealthily climbing, while the male exerts himself lovingly to bring his mate those delightful things in which the tropical forest is rich—fruits above all, but occasionally a delicate mouse or juicy frog. He flies with his booty to the tree and gives a peculiar knock, which his mate knows as his signal, and thrusts her beak through the narrow window, welcoming her meal." At the end of the period of incubation, C. M. Woodford says, "the devoted husband is worn to a skeleton."

But animals, like men, have their vices, and birds, generally so ideal in their behaviour, are sometimes criminals. Ornithologists assure us that the degree of parental care varies not only in nearly related species, but also among members of the same species. We need not lay much stress on the fact that a bird occasionally slips its egg into a neighbour's nest, for when a partridge thus uses a pheasant's rough bed, or a gull that of an eider-duck, it is likely enough that the intruder had been disturbed from her own resting-place when about to lay. We approach something different in the case of the American Ostrich (*Rhea*), the female of which is quite ready to utilise a neighbour's burrow; nor does the owner seem to object, for all the brooding is discharged by the male, "and it is no great art to be patient and magnanimous at another's expense." Again, in the case of the American Ani (*Crotophaga ani*), of whose habits we unfortunately know little, a number of females sometimes lay their eggs in a common nest.

We are so glad to hear the cuckoo's call in spring that we almost forget the wickedness of the voluble bird. The poets have helped us, for they have generously idealised, in fact idolised, the cuckoo, the "darling of the spring," "a wandering voice babbling of sunshine and of flowers," a "sweet," nay more, a "blessed bird." But the cuckoos have hoaxed the poets, for they are even worse than their legendary reputation of being sparrow-hawks in disguise;

they are "greedy feeders," says Brehm, "discontented, ill-conditioned, passionate fellows; in short, decidedly unamiable birds." The truth must be told, the cuckoo is an immoral vagabond, an Ishmaelite, an individualist, a keeper of game "preserves." There are so many males that they have perverted and thoroughly demoralised the females; there is no true pairing; they are polyandrous. The birds are too hungry for genuine love, though there is no lack of passion; while by voraciously devouring hairy caterpillars they have acquired a gizzard-fretting feltwork in their stomachs, and for all I know are cursed by dyspepsia as well as by a constitutionally evil character. It is not quite correct to say that the cuckoo-mother is immoral because she shirks the duties of maternity; it is rather that she puts her young out to nurse because she is immoral.[1] The so-called "parasitic" trick is an outcrop of an egoistic constitution which shows itself in many different ways. The young bird, "a dog in the manger by birth," evicts the helpless rightful tenants whether they are still passive in the eggs or more assertive as nestlings, and as he grows up a spoilt child his foster parents lead no easy life. But though the poets have been hoaxed, I do not believe that the nurses of the fledgling are; it seems rather as if the naughtiness of their changeling had some charm.

Of course there is another way of looking at the cuckoo's crime. It is advantageous, and there is much art in the well-executed trick by which the mother foists her several eggs, at intervals of several days, into the nests of various birds, which are usually insectivorous and suited for the upbringing of the intruder. I think there is at least some deliberation in this so-called instinct. Nor should one forget that the mother occasionally returns to the natural habit of hatching her own eggs,—a pleasant fact which several trustworthy observers have thoroughly established. Still, in spite of the poets, the note of this "blessed bird" must be regarded as suggestive of sin!

[1] The student will notice that I have occasionally used words which are not strictly accurate. I may therefore say definitely that I do not believe that we are warranted in crediting animals with moral, æsthetic, or, indeed, any conceptions.

There is much to be said about the domestic life of animals—their courtship, their helpful partnership, and their parentage—but perhaps I have said enough to induce you to think about these things more carefully. Many of the deepest problems of biology—the origin and evolution of sex, the relation of reproduction to the individual and to the species—should be considered by those who feel themselves naturally inclined to such inquiries; moreover, in connection with our own lives, it is profitable to investigate among animals the different grades of the love of mates, and the relation between the rate of reproduction and the degree of development. First, however, it were better that we should watch the ways of animals and seek after some sympathy with them, that we may respect their love, and salute them not with stone or bullet, but with the praise of gladdened eyes.

Ruskin's translation of what Socrates said in regard to the halcyon is suggestive of the mood in which we should consider these things.

Chærophon. "And is that indeed the halcyon's cry? I never heard it yet; and in truth it is very pitiful. How large is the bird, Socrates?"

Socrates. "Not great; but it has received great honour from the gods, because of its lovingness; for while it is making its nest, all the world has the happy days which we call halcyonidæ, excelling all others in their calmness, though in the midst of storm."

"We being altogether mortal and mean, and neither able to see clearly great things nor small, and for the most part being unable to help ourselves even in our own calamities, what can we have to say about the powers of the immortals, either over halcyons or nightingales? But the fame of fable, such as our fathers gave it to us, this to my children, O thou bird singing of sorrow, I will deliver concerning thy hymns; and I myself will sing often of this religious and human love of thine, and of the honour thou hast for it from the gods."

Chærophon. "It is rightly due indeed, O Socrates, for there is a twofold comfort in this, both for men and women, in their relations with each other."

Socrates. "Shall we not then salute the halcyon, and so go back to the city by the sands, for it is time?"

CHAPTER VII

THE INDUSTRIES OF ANIMALS

1. *Hunting*—2. *Shepherding*—3. *Storing*—4. *Making of Homes*
5. *Movements*

It is likely that primitive man fed almost wholly upon fruits. His early struggles with animals were defensive rather than aggressive, though with growing strength he would become able for more than parrying. We can fancy how a band of men who had pursued and slain some ravaging wild beast would satisfy at once hunger and rage by eating the warm flesh. Somehow, we know, hunting became an habitual art. We can also fancy how hunters who had slain a mother animal kept her young alive and reared them. In this or in some other way the custom of domesticating animals began, and men became shepherds. And as the hunter's pursuits were partially replaced by pastoral life, so the latter became in some regions accessory to the labours of agriculture, with the development of which we may reasonably associate the foundation of stable homesteads. Around these primary occupations arose the various human industries, with division of labour between man and woman, and between man and man.

These human industries suggest a convenient arrangement for those practised by animals. For here again there are hunters and fishers—beasts of prey of all kinds—pursuing the chase with diverse degrees of art; shepherds, too, for some ants use the aphides as cows; and farmers without doubt, if we use the word in a sense wide enough to include those who collect, modify, and store the various fruits of the earth.

In illustrating these industries, I shall follow a charming volume by Frédéric Houssay, *Les Industries des Animaux*, Paris, 1890.

1. **Hunting.**—Of this primary activity there are many kinds. The crocodile lies in wait by the water's edge, the python hangs like a lian from the tree, the octopus lurks in a nook among the rocks, and the ant-lion (*Myrmeleon*) digs in the sand a pitfall for unwary insects. The angler-fish (*Lophius piscatorius*) is somewhat protectively coloured as he lies on the sand among the seaweeds; on his back three filaments dangle, and possibly suggest worms to curious little fishes, which, venturing near, are engulfed by the angler's horrid maw, and firmly gripped by jaws with backward-bending teeth. Many animals prowl about in search of easy prey—eggs of birds, sleeping beasts, and small creatures like white ants; others would be burglars, like the Death's Head Moth (*Sphinx atropos*) who seeks to slink into the homes of the bees; others are full of wiles, witness the cunning fox and the wide-awake crow. Many, however, are hunters by open profession, notably the carnivorous birds and mammals. If these hunters could speak we should hear of many strange exploits; such, for instance, as that of a large spider which landed a small fish. The ins and outs of their ways are most interesting, especially to the student of comparative intelligence. Think of the Indian *Toxotes*, a fish which squirts drops of water on insects and brings them down most effectively; several birds which let shells drop from a height, *e.g.* the Greek eagle (*Gypaëtos barbatus*), which killed Æschylus by letting a tortoise drop on his head; the grey-shrike (*Lanius excubitor*), which spikes its victims on thorns; and, strangest perhaps, the slave-making expeditions of the Amazon ants. All strength and wiles notwithstanding, the chase is often by no means easy; the hare grows swift as well as the fox, many grow cautious like trout in a much-fished stream, scouts and sentinels

are often utilised, the weak combine against the strong, and the victims of even the strong carnivores often show fight valiantly.

2. **Shepherding.**—Although the ants are the only animals which show a pastoral habit in any perfection, and that only in four or five species (*e.g. Lasius niger* and *Lasius brunneus*), I think that the fact is one about which we may profitably exercise our minds. I shall follow Espinas's admirable discussion of the subject.

We may begin with the simple association of ants and aphides as commensals eating at the same bountiful table. But as ants discovered that the aphides were overflowing with sweetness, they formed the habit of licking them, the aphides submitting with passive enjoyment. Moreover, as the ants nesting near the foot of a tree covered with aphides would resent that others should invade their preserves, it is not surprising to find that they should continue their earthen tunnels up the stem and branches, and should eventually build an aerial stable for some of their cattle. Thither also they transport some of their own larvæ to be sunned, and as they carried these back again when the rain fell, they would surely not require the assistance of an abstract idea to prompt them to take some aphides also downstairs. Or perhaps it is enough to suppose that the aphides, by no means objecting to the ants' attentions, did not require any coaxing to descend the tunnels, and eventually to live in the cellars of the nests, where they feed comfortably on roots, and are sheltered from the bad weather of autumn. In autumn the aphides lay eggs in the cellars to which they have been brought by force or coaxing or otherwise, and these eggs the ants take care of, putting them in safe cradles, licking them as tenderly as they do those of their own kind. Thus the domestication of aphides by ants is completed.

Now what is the theory of this shepherding? (1) We have no warrant for saying that the ants have deliberately domesticated these aphides, as men have occasionally added to the number of their domesticated animals. It does not seem to me probable that even primitive man was very deliberate in the steps which led to the first domestications. (2) Nor is it likely that the process began in a casual way, and that it became predominant in four or five species in the course of natural selection. For the habit is more a luxury than a necessity, and it is not likely to have been evolved before the establishment of the sterile caste of workers, who have no means of transmitting their experience. Moreover, initial steps are always difficult to explain on this theory. (3) The theory which seems to me warrantable is that the habit arose by a gradual extension of habits previously established, that it was neither deliberate nor casual in its origin, but a natural growth, beginning neither in a clever experiment nor in a fortunate mistake of an individual worker ant, but the outcome of the community's progressive development in "intellectual somnambulism," helped in some measure by the sluggish habits of the aphides. And, if you wish, the formula may be added, "which was justified in the course of natural selection."

3. **Storing.**—Not a few animals hide their prey or their gatherings, and with marvellous memory for localities return to them after a short time. But genuine storing for a more distant future is illustrated by the squirrels, which hide their treasures like misers. Many mice and other rodents do likewise, and in some cases the habit seems to become a sort of craze, so large are the supplies laid in against the winter's scarcity. Very quaint are the sacred scarabees (*Ateucus sacer*), which roll balls of dung to their holes, and sometimes collect supplies at which they gnaw for a couple of weeks. Some ants (*e.g. Atta barbara*) accumulate stores of grain, occasionally large enough to be worth robbing; and there is no doubt that they are able to keep the seeds from germinating for a considerable time, while they stop the germination after it has begun by gnawing off plumule and radicle and drying the seeds afresh. Dr. M'Cook's account of the agricultural ant of Texas (*Pogomyrmex barbatus*) gives even more marvellous illustrations of farming habits, for these ants to a certain extent at least cultivate in front of their nests a kind of grass with a rice-

like seed. They cut off all other plants from their fields, and thus their crops flourish.

But animals store for their offspring as well as for themselves. The habit is very characteristic of insects, and is the more interesting because the parents in many cases do not survive to see the rewards of their industry. Sometimes, indeed, there is no industry, for the stores of other insects may be utilised. Thus a little beetle (*Sitaris muralis*) enters the nest of a bee (*Anthophora pilifera*) and lays its eggs in the cells full of honey. More laudable are the burying-beetles (*Necrophorus*), which unite in harmonious labour to bury the body of a mouse or a bird, which serves as a resting-place for their eggs and as a larder for the larvæ. The *Sphex* wasp makes burrows, in which there are many chambers. Each chamber contains an egg, and is also a larder, in which three or four crickets or other insects, paralysed by a sting in the nervous system, remain alive as fresh meat for the Sphex larva when that is hatched. After the *Sphex* has caught and stung its cricket and brought it to the burrow, it enters at first alone, apparently to see if all is right within. That this is thoroughly habitual is evident from Fabre's experiment. While the *Sphex* was in the burrow, he stole away the paralysed cricket, and restored it after a little; yet the wasp always reconnoitred afresh, though the trick was played forty times in succession. Yet when he substituted an unparalysed cricket for the paralysed one, the *Sphex* did not at once perceive what was amiss, but soon awoke to the gravity of the situation, and made a fierce onslaught on the recalcitrant victim. So it is not wholly the slave of habit.

4. **Making of Homes.**—Houssay arranges the dwellings of animals in three sets—(*a*) those which are hollowed out in the earth or in wood; (*b*) those which are constructed of light materials often woven together; and (*c*) those which are built of clay or similar material. We may compare these to the caves, wigwams, and buildings in which men find homes.

Burrows are simplest, but they may be complex in

details. Those of the land-crabs (*Gecarcinus*), the wood-cutting bees (*Xylocopa*), the sand-martens, the marmots, the rabbit, the prairie dogs, illustrate this kind of dwelling in various degrees of perfection.

The male stickleback (*Gasterosteus*) weaves and glues

FIG. 27.—Swallows (*Chelidonaria urbica*) and their nest. (After Brehm.)

the leaves and stems of water-plants; the minutest mouse (*Mus minutus*) twines the leaves of rushes together; the squirrel makes a rougher nest; the orang-utan and the chimpanzee construct shelters among the branches; but the nests of many birds are by far the most perfect works of animal art.

Of buildings, the swallows' nests by the window, and the

paper houses which wasps construct, are well known; but we should not forget the architecture of the mason-bees, the great towers of the termites, and the lodges of the beavers.

Perhaps I may be allowed to notice once again, what I have suggested in another chapter, that while many of the shelters which animals make are for the young rather than for the adults, the line of definition is not strict, and some which were nests to begin with have expanded into homes —an instance of a kind of evolution which is recognisable in many other cases.

5. **Movements.**—But animals are active in other ways. All their ways of moving should be considered—the marvel-

FIG. 28.—Flight of crested heron, ten images per second. (From Chambers's *Encyclop.*; after Marey.)

lous flight of birds and insects, the power of swimming and diving, the strange motion of serpents, the leap, the heavy tread, the swift gallop of Mammals. All their gambolings and playful frolics, their travels in search of food, and their migrations over land and sea, should be reckoned up.

Most marvellous is the winged flight of birds. As a boat is borne along when the wind fills the sails, or when the oars strike the water, and as a swimmer beats the water with his hands, so the bird beating the air backwards with its wings is borne onward in swift flight. But the air is not so resistent as the water, and no bird can float in the air as a boat floats in the water. Thus the stroke has a downward as well as a backward direction. When there is more of the downward direction the bird rises, when there is more of the backward direction it speeds forward; but usually the stroke is both downwards and backwards, for the lightest bird has to keep itself from falling as it flies. The hollowness and sponginess of many of the bones combine strength of material with lightness, and the balloon-like air-sacs connected with the lungs perhaps help the birds in rising from the ground; but, buoyant as many birds are, all have to keep themselves up by an effort. But the possibility of flight also depends upon the fact that the raising of the wing in preparation for each stroke can be accomplished with very little effort; the whole wing and its individual feathers are adjusted to present a maximum surface during the down-stroke, a minimum surface during the elevation of the wing. There are many different kinds of flight, which require special explanation—the fluttering of humming-birds, the soaring of the lark, the masterful hovering of the kestrel, the sailing of the albatross. The effortless sailing motion of many birds is comparable to that of a kite, "the weight of the bird corresponding to the tail of the kite;" it is possible only when there is wind or when great velocity has been previously attained.

FIG. 29.—From St. John's *Wild Sports*.

PART II

THE POWERS OF LIFE

CHAPTER VIII

VITALITY

1. *The Task of Physiology*—2. *The Seat of Life*—3. *The Energy of Life*—4. *Cells, the Elements of Life*—5. *The Machinery of Life*—6. *Protoplasm*—7. *The Chemical Elements of Life*—8. *Growth*—9. *Origin of Life*

1. **The Task of Physiology.** — So far we have been considering the ways of living creatures, as they live and move, feed and grow, love and fight; as they build their homes and tend their young. We shall now turn to the inner mysteries, and seek, so far as we may, to fathom the wisdom of the hidden parts. We shall describe the machinery—the means by which the forces of life cause those movements by which we recognise their presence.

This study is called physiology; and the plan of our sketch of present knowledge will be as follows: We shall first try to realise what we mean by life; we shall then limit ourselves to the consideration of certain kinds of life, and attempt to make plain in what parts of all living creatures are the forces of life most active. Having done this, we shall describe the life processes of the simplest creatures, and then those of the higher animals.

It is not easy to say clearly what we mean by life; but we recognise as one of its characteristics the power of movement.

Still, this gives no distinction between the blowing of wind and the life of man; but the other characteristics of life will be realised as we proceed in our analysis; it is certain that without movement there is no life. Further thought may lead us to define life as that "complex of forces which produces form." Thus the star-like crystals of a snowflake, the diamond drops of dew, the overshadowing mountains, would all be imaged in our minds as living, though of more lowly life than the lichens of the bare hill-tops, the grass of the plains, or man himself. We have no space here to trace the connections between such an idea and the beliefs of all simple peoples, and the inspirations of all poets, but the similarity is evident, and the usefulness in philosophy of such generalised conceptions is great.

But the physiology which we shall sketch here will be a narrower one; it will be confined to the life of plants and animals, and we shall attempt to show precisely how that life is separated from the life of the dust and of the air.

2. **The Seat of Life.**—Now in what parts within the living body are the life forces most actively at work?

When we look at any living creature we are all too willing, even if the wonder of life stirs within us, to remain satisfied with a vague apprehension of a mystery. It is strange that so many generations of men passed away before any steps were taken towards a conception of the intimate material processes of life and growth and death. The moving train has been watched, but the engine and the stoker have been almost unnoticed.

Let us consider the growth of a tree. The outward manner of its growth we can observe, a few superficial details of its inner life we already know, and of this knowledge we may look for great expansion, but the ultimate processes of its life are still a complete mystery to us.

The tree is alive, but is it all alive? Cut a stake from its heart and plant it in the ground; it will not grow, and shows no signs of life, but we are not puzzled; the tree, we think, can only live as a whole, and we know how easily most living things are killed by local injuries. But if we cut a stake from the outer part of the tree, leaving the bark on, and set it in the ground, it may happen that buds will appear, pushing through the bark, and stretching out into shoots.

There is a mystery for us to begin with: some parts of a tree may have a life of their own. Indeed, we all know that gardeners do not rear geraniums and other plants from seeds, but from cuttings. Potatoes, as we know, will give origin to new plants, and even small parts of potatoes will do so. Roses are grafted into the stems of the wild brier, and in this way two life-currents are mingled. We may remember, too, that all seeds are only parts that have become separated from the parent plant. We ourselves, formed in the darkness of the womb, were separated at birth from the mothers who bore us.

Let us think of the seeds of plants for a little. Formed in the warmth and brightness of the summer sun, ripened in the glow of autumn, they fall to the ground, are carried hither and thither by trickling runlets of water, by the winds, by animals, and scattered over new pastures. Through the long chill of winter they remain asleep; but not dead,—slow preparation is being made for the new day. With the warm winds of spring—when the birds come back to us and sing their first songs of love and courtship—the countless buds of the woods, the gardens, and the hedgerows, all the seeds we sowed in the autumn, all the corn we scattered in the first hours of the new morning, awake; the buds burst, the tiny leaves unroll; in the seeds there is a great activity,—the slender shoots stretch forth—spring passes into summer—and we await the harvest.

3. **The Energy of Life.**—What is the cause of this strength of life? How is it that in an acre of forest tons of solid matter are lifted high into the air, while the branches waving under the blue sky seem to enjoy the brightness of the sun after the gloom of winter? This assertion of the poets of the gladness of nature at the springtime is no mere wandering fancy, it is simple truth; the intensity of life at that time is due entirely to the greater warmth of the air

and increased brightness of the sun; and since there is no reason why we should not believe in a simple order of consciousness in all simple creatures, that consciousness would certainly become gladness with increased vigour of life, just as it is with ourselves.

The energy of life, we say, is due to the energy reaching us from the sun; but how is the radiant energy of the ether used to place the growing shoots on the forest tops, and how is it transformed into the potential energy of wood and other substances? Our trains move by virtue of the energy stored in the coal; we burn that in a certain place and get expansive energy of steam, which by mechanical arrangements we convert into the moving energy of the engine; but where is the boiler, and where the machinery in the plant, by which the energy of the sun's rays is transmuted into life? The partial answer to this riddle has been found. If we break a newly-grown branch we find the tissues filled with a watery slimy sap. If we open a bud we find the slimy stuff again; under the bark of trees we find it once more; it is within the tissues of a bulb, and in growing seed; indeed, in all those parts of a plant which are capable of independent life we always find what we may call for the moment this slimy sap; while in the hard inner parts of the tree, which we know can live no more, we find nothing of the kind,—such tissues are quite empty. Life, therefore, we find to have something to do with a certain substance, and this is the first step towards understanding the machinery we have spoken of; we have, as it were, found that the movement of the train is due to the engine, but we do not understand that engine.

4. **Cells, the Elements of Life.**—Let us leave the trees now for a little, and turn to the simplest of all living creatures, which live in water and in damp places. They are so small that only a few of the larger ones can be seen as tiny specks moving about in the water in which they live. But they can be seen quite easily with a microscope. We find them to be little transparent drops of living matter. They are not really drops; many of them have

distinct shapes, others constantly change their form. They move, indeed, by a kind of flowing; one part of their body is pushed out and a part on the farther side drawn in. Some of these lowly creatures have skeletons or shells of lime or of flint. Great numbers of these shells, when the little inmates are dead, form beds of chalk and ooze. Now all living creatures begin life in this way; at first they are tiny masses of a jelly-like translucent stuff. Each mass gets a skin or surrounding wall; if fed, it grows larger, and a wall is built up inside it, making its house a two-roomed one. This process goes on and on; the whole mass grows larger and larger, and becomes divided up into a corresponding number of compartments. The chambers are not quite separated; there are always holes left in the walls, through which strands of the jelly-like stuff pass, and so all of them are connected. The divisions in each separate kind of animal or plant take place in a special way, until at last the whole body is built up, with all its peculiarities of form and internal arrangement. The cells of an animal's body do not, however, form walls as definitely as do those of plants.

In an ordinary plant there are millions of those compartments; they are called cells, from their likeness in general appearance to the cells of a honeycomb; and the enclosed stuff that we have spoken of as jelly is called protoplasm, because it is believed that the first living things that were formed were little drops of jelly-like stuff, not unlike that within the cells, or composing the animalculæ in water. Protoplasm, wherever it occurs, from the highest to the lowest forms of life, is supposed to have, within certain limits, a similarity of nature.

In some plants the cells are large enough to be visible to the naked eye, but the cells of most plants and animals are so small that they can only be seen with a microscope.

We can now give a complete answer to the question, What parts of a tree are alive? It is only the protoplasm of the cells. The walls of the cells are more or less dead. As the cells grow older and larger, and the walls become thicker, the amount of protoplasm within gets relatively

K

less; at last it slowly dies and withers away; the cells are left empty; and that is why the stake cut from the old hard part in the middle of the tree could not grow,—it was quite dead.

5. **The Machinery of Life.**—We have found that, in some way, the protoplasm within the cells is the machinery of life. For simplicity, we shall speak of protoplasm as living matter. This living matter in plants is such that it can transform the energy of sunlight into potential energy of complicated substances such as wood. This transformation of energy is one of the chief labours of plants in the world. A great deal of the energy that reaches their living matter is used for their own upward growth; so that, as we said before, thousands of tons of matter are every year, over every acre of forest, raised high into the air.

In animals the living machinery is in certain ways of a different nature. An animal eats the substances made by the plant; the potential energy stored in these is used by it in moving about, and so transformed into energy of motion. The life of plants is chiefly shown in the storage of energy, the life of animals in the use of that store. Chiefly we say, for plants also move to a slight extent; as a whole, when they twine around a tree or bend towards the sun; and in their parts when the sap rises and falls. Animals also, to a slight extent, build up substances of high potential energy.

So far all is certain, but when we inquire by what arrangement of parts the living matter is able to be a machine for the transformation of energy, we are unable to form any conception. Soon after the discovery of the cells and their living contents, certain philosophers, who must have very faintly realised the necessary physical conditions, arguing from the analogy of machines as men construct them, supposed that the activities of the living machinery could be deduced from the structural arrangements of the cells; they supposed that the living matter, a part within it called the nucleus, and the cell-wall, were in themselves the parts of a machine, and that the various activities of the cells were due to varying shapes of wall, and disposition of its visible parts. It was soon shown, however, that the wall was not a necessary part of the living matter, and that the nucleus did not always occur. The cell is a machine, not in virtue of the disposition of its visible parts, but as a consequence of the arrangement of its molecules. We know this much about the living machinery, that it is far more perfect than the machinery of our steam-engines, the perfection of a machine being measured by the relation between the energy which enters it and that which leaves it as work done.

"Joule pointed out that not only does an animal much more nearly resemble in its functions an electro-magnetic engine than it resembles a steam-engine, but also that it is a much more efficient engine; that is to say, an animal, for the same amount of potential energy of food or fuel supplied to it—call it fuel to compare it with other engines —gives you a larger amount converted into work than any engine which we can construct physically." And Joly has expressed the contrast between an inanimate material system and an organism as follows: "While the transfer of energy into any inanimate material system is attended by effects retardative to the transfer and conducive to dissipation, the transfer of energy into any animate material system is attended by effects conducive to the transfer and retardative of dissipation."

It is from protoplasm that we must start in our study of living machinery; let us see how far we can attain to exact conceptions of its nature. We will first describe shortly what is known as to the structure of protoplasm or living matter, chiefly to show how hopeless is any attempt at a solution of the problem in terms of visible structure. The powers of the microscope are limited by the physical nature of light, and that limit has already nearly been reached; and yet we know the structure of matter is so excessively minute that within the compass of the finest fibre visible with the microscope there is room for the most intricate structural arrangements.

6. **Protoplasm.**—Protoplasm used commonly to be described as a structureless mass; we now know that it often has a structure somewhat like a heap of network. It is a complex of finely-arranged strands, with knots or swellings

at the junctions of the strands, and with, in each cell, one or more central and larger swellings, probably of a highly specialised nature, called nuclei. The size of the meshes varies, and they are filled "now with a fluid, now with a more solid substance, or with a finer and more delicate network, minute particles or granules of variable size being sometimes lodged in the open meshes, sometimes deposited

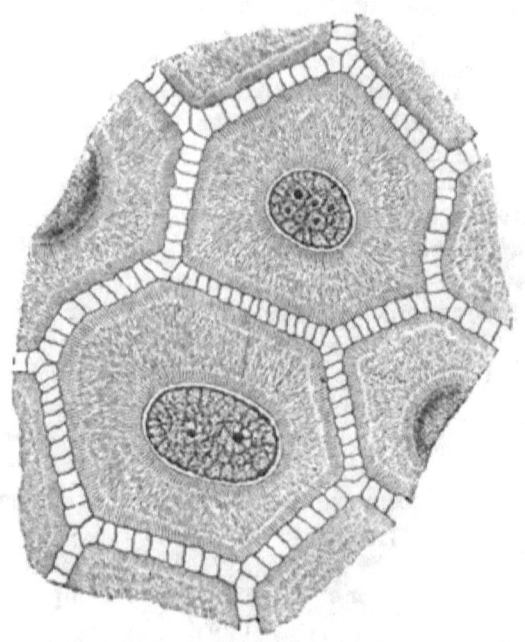

FIG. 30.—Adjacent animal cells showing the nucleus, the protoplasmic network, and the bridges vitally uniting cell and cell across the intervening intercellular substance. (From the *Evolution of Sex*; after Pfitzner.)

in the strands of the network. Sometimes, however, the network is so close, or the meshes filled up with material so identical in refractive power with the bars or films of the network, that the whole substance appears homogeneous." The only means we have of getting any further knowledge of this arrangement is by staining it with various dyes, and observing the effects of the dyes upon the various parts. "Analysis with various staining and other reagents leads to the conclusion that the substance of the network is of a different character from the substance filling up the meshes. Similar analysis shows that at times the bars or films of the network are not homogeneous, but composed of different kinds of stuff; yet even in these cases it is difficult, if not impossible, to recognise any definite relation of the components to each other such as might deserve the name of structure." Plainly there is not much light to be got by further investigations in this direction. Ordinary chemical analysis, too, is of little avail; for how can we say what parts of the mass are alive, even if we could separate part from part? Is it only the meshwork that is really living matter, or are the granules part of it, or are the fluid contents the chiefly vital substance?

Let us turn now to the activities of the living stuff, and see what we can learn from them. We have already spoken of one of the activities of living matter, especially of plant protoplasm, that of surrounding itself with a wall. Now we might at first be inclined to suppose that the wall was simply due to a hardening and drying of the soft substance at those places where it touched the air. It is possible that that may have been the stimulus which caused, as a reaction, the wall-making at the dawn of life, and which may still have some connection with it; but what we have to take note of is the fact that the walls, as they are made by the higher plants, have always a definite structure and chemical nature.

If we examine the cells of the leaves of a plant growing in the sunlight, we find the green colouring matter to be generally collected in little rounded masses. Looking more closely, we find it to be the fluid which fills the meshwork of the masses. At certain points in the meshwork we find minute masses of starch constantly being formed. They seem to pass along the strands and collect in the centre of the network, until quite a large mass of starch is accumulated there. If we examine the plant at night, some time after darkness has set in, we find no traces of starch in the cells of the leaves. There is evidence that the starch has been transformed into sugar, and can then, by osmotic and perhaps by other processes, be removed from the leaves, and carried by the vessels to all parts of the plant. So we get a first notion of how a plant is fed. Starch is a compound containing carbon and the elements of water. The carbon, we know, comes from the carbonic acid of the air; the water is absorbed by the roots from the soil. In some way the living matter of the cells, by means partly of the green colouring matter, is able to transform the energy of the sun's rays into potential energy of a combustible substance—starch; so we get clear evidence of a machinery for the transformation of energy. We have taken a plant as our example throughout, partly because the cells are more evident than in animals, and partly because the chemical processes give evidence of a transformation of kinetic into potential energy more clearly than do those of animals; for the animals eat the plants, and so by using the potential energy of plant substance are able to live and move.

We have now some idea of the sources of the energy of life. Plants get their food from the air by their leaves, and from the soil by their roots, which absorb water and salts dissolved in water. By aid of the energy of sunlight they build these up into complex substances, which they use for the growth of their living matter, for the formation of supporting structures, and for other purposes. Animals eat these substances. They build up their own bodies of living matter and supporting structures, and they move about.

In order to get a clearer notion of the nature of living matter we must attempt to trace the manner in which these various substances are built up. We have first to discover what are the substances that are made. In all living creatures there are, in addition to water and salts, such as common salt and soda, three groups of stuffs :—

Carbohydrates, such as starch and sugar, made of carbon with hydrogen and oxygen in the same proportions as they occur in water;

Fats—substances containing the same three elements, but with a smaller proportion of oxygen;

Proteids—substances containing always carbon, hydrogen, oxygen, and nitrogen, with a small percentage of sulphur.

The constitution of proteids is difficult to determine.

The above elements are always present, and in proportions which vary within narrow limits; but in addition to these substances there seem to be always present others which, when the proteids are burnt, remain as ash in the form of salts chiefly chlorides of sodium and potassium, but also small quantities of calcium, magnesium, and iron, as chlorides, phosphates, sulphates, and carbonates. The molecule of a proteid must be very complex; thus that of albumen is, at its smallest, C_{292}, H_{481}, N_{90}, O_{83}, S_2; most probably it is some multiple of this.

The food-stuffs of plants, then, are salts, water, and carbonic acid, and a certain amount of oxygen. Of these, by means of the sun's energy, they build up complex substances —carbohydrates, fats, proteids, which, with salts, water, and oxygen, serve as the food of animals. The living matter, the machinery by which all this is done, is, if it can be classed at all, a proteid. But this only means that all living matter contains the five essential elements and some others which in the ash exist as salts.

The various services which the different food-materials are set to within the body will be described later, when we are considering the details of the animal economy. Here we shall take note of the elements that enter into the construction of the food-stuffs.

7. **The Chemical Elements of Life.**—There are sixty-eight elements to be found in varying abundance upon the earth, but by analysis of the food-stuffs and of living matter itself, we find that only twelve of these occur with any constancy in organisms. They are carbon, hydrogen, oxygen, nitrogen, sulphur, phosphorus, chlorine, potassium, sodium, calcium, magnesium, and iron.

Now nine of these elements form sixty-four per cent by weight of the earth's crust; while aluminium and silicon, substances that are only very occasionally found in living creatures, form thirty-five per cent, being the chief constituents of quartz and felspar, sand and clay, in short the greater part of all rocks. All the other elements, three of which—hydrogen, nitrogen, and phosphorus—enter into life, form the remaining one per cent.

Since the ultimate analysis of the objective side of life seems to show that life is to be pictured as matter in an unstable and constantly altering condition, it will be of interest to find the conditions that determine which of the elements are to take part in it.

It seems that matter in order to enter into life must be—

(1) Common, (2) mobile, that is capable of easily entering into solution or becoming gaseous, and (3) capable of forming many combinations with other elements.

Nine of the elements fulfil the first condition; a tenth, nitrogen, is the chief constituent of the atmosphere; while hydrogen is present everywhere in water. Why do not aluminium and silicon take their share in life? Because they do not fulfil conditions (2) and (3). Their oxides are quartz and aluminia, two of the hardest substances known. Emery and ruby are two forms of aluminia; while the oxide of carbon, the source of all the carbon used in life, is a gas.

Carbon, which takes so great a part in life processes that the chemistry of organic substances is commonly spoken of as the chemistry of the compounds of carbon, fulfils all three requirements in an eminent degree. For although in its pure form a solid, and sometimes a very hard substance, yet it readily forms an oxide which is present in the atmosphere, and, as we know, serves as one of the chief foods of plants. Its power of entering into combination with other elements is practically infinite. Nitrogen, although by itself an inert form of matter, is able to combine with carbon compounds and add fresh complexity.

It is easy to see why water is so important in life. It dissolves the other substances, and so allows them to come into closer contact, and to change in position more easily, than if they were solid. So the first stuff that was complex and unstable enough to be properly described as living was almost certainly formed in water, long ago, when the conditions of greater heat, and consequently greater mobility of all substances, made chemical changes more active.

The importance of the solvent power of water in a complex organism is obvious when we think of the blood, the great food stream and drain. It is shown in an interesting

Vitality

way by the suspended animation of a dried seed, which will remain for years dormant, but ready when moistened to spring into active life.

How, then, are these substances built up into living creatures? Let us, that we may see this matter clearly, think for a moment of the conditions of life of the simplest creatures, the formless masses of living matter. All that the simplest plants need is water holding oxygen, carbonic acid, and salts in solution. Out of these simple materials, by the magic touch of their living bodies, they can build up

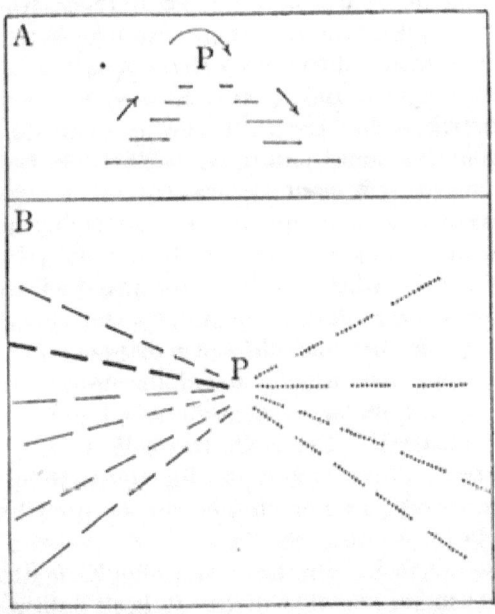

the complex matter of which those bodies are made; so that they can grow and divide until there may be hundreds in place of a single one. The image we must form of this increase of living matter is that step by step substances of an ever-growing complexity are made, one from the other, until at last a substance so unstable is made that it begins to break down into simpler forms of matter at the least deviation from the precise conditions in which it was made, and perhaps also with a ferment-like action causes changes in all that it touches; and this we call the living matter. As this living matter breaks down into simpler substances, or as it causes surrounding substances to break down, energy is set free for use in movement. We may make a diagram of this process. The steps go up and down; the top one we call protoplasm (Fig. A). This shows only one line of ascent and one of descent. There may be many, all going on at the same time, as is shown in Fig. B. But these are much too simple; they show continuous ascending and descending stairs, but each step does not really result directly from the one below, but must result from two or more stairs meeting; and at each meeting there must be substances formed which are useless, and begin to break down, or are cast out of the system at once.

8. **Growth.**—The power of growth, of adding to itself substance of the same nature as itself, is the real mystery of living matter. A crystal grows out of its solution, the star or pyramid is built up with perfect regularity, but the process is much simpler than the growth of living matter. The substance of which it will be formed is already there, but the protoplasm has to make its own substance as it grows. This is the true difference between the two processes, and not, as is usually stated, that a crystal grows by depositing matter on its surface, while a cell grows by putting matter within itself. For when two cells fuse, which often occurs, growth is really as much by aggregation as in the case of a crystal, and such manner of growth is made possible simply because the two cells are masses of matter of equal complexity. But when less complex matter is given to a cell it cannot add that matter to itself until it has been transformed into substance as complex as itself. This change can only be effected within the little laboratory of the cell itself. The fact that the growth of a crystal may be endless, while that of a cell is limited, which is usually cited as the distinctive difference, is a consequence of the necessity of the protoplasm for forming its own substance within its own substance. For when a cell grows in size the ratio of its surface to its volume constantly decreases; and therefore, since new material can only be absorbed

through the surface, there must be a certain size of cell at which the rate of absorption is just sufficient for the nourishment of the protoplasm. Beyond this point a cell cannot grow; but if it divides, then the mass to be fed remains the same, while the absorbing surface is increased. This, then, is the necessitating cause of cell-division. But it would be unwise to suppose that there are not other causes that help to produce this result, which has as a consequence the possibility of immense variety of disposition of the daughter cells, and therefore of organic forms; for, to begin with, a more obvious means of obtaining increased surface would be for the cell merely to become flattened or to spread out irregularly, which, indeed, we see in many of the Protozoa.

Since their growth implies cell-division as one of its consequences, and since cell-division is the basis of reproduction, synonymous, indeed, in the Protozoa with reproduction, we get the idea of successive generations of animals as merely the continued growth of former generations. This makes intelligible to us all the facts of heredity which are so surprising if we conceive of each generation as a number of untried souls that have left some former dwelling-place to come and live among us. Our children are, in truth, absolutely portions of ourselves. If this be so, we must imagine in the ovum—the tiny mass of protoplasm from which we are formed by continued division—a most extraordinary subtlety of constitution.

Try to picture the complexity of the arrangement of parts. There are two tiny masses of protoplasm; so far as we can see they are the same, yet from one will grow a man, from the other a tree. If the germ that will grow into a human being could only properly be fed outside the body of the mother, so far as we know it might leave that body as an almost invisible cell, and would grow and divide, add cell to cell, until the creature was fully formed—sculptured out of dust and air. Our early life within the womb, our nourishment by the blood of our mother, is only nature's way of preserving us from injury. What we shall be is already marked out before the egg begins to grow.

It is only the highest animals who are thus shielded.

The birds cover their eggs with their wings. The butterfly lays hers where the grubs will find their food. The star-fish cast theirs adrift in the sea. The same story is true of the flowers. They are the nursing mothers. In their heart the young plant grows until the first leaves appear; not till then does it drop away, and not without food prepared and placed ready for use — enclosed in what we call the seed. But the seaweed, like the star-fish that crawls upon it, allows its young seeds quite unformed to be floated away by the tide.

All seeds, then, are parts of the living matter of the parent; some leave naked and without food; others are protected by shells or by husks which are filled with food; others live within the mother until they have ceased really to be seeds, and are fully formed new creatures.

Now the living matter of any simple organism is so much the same throughout the whole body that almost any part of it will do to build the new generation from. Thus, although the sea-anemone does sometimes set apart certain cells as seeds, yet any part of the body will, if cut off, grow into a complete creature. The same thing is true of a moss plant. But the more highly organised animals have their living matter set to such different service in their various organs that most of it does not keep all the qualities of the whole creature, but only of that part to which it belongs. Thus if a starfish lose its arm, another will grow from the stump. A snail can in this way repeatedly regrow its horns. Even so highly developed an animal as a lizard can grow a new tail. With ourselves this power is confined to growing new skin if we lose part of it, or mending a bone if it be broken, and other similar processes.

When we clearly understand in what way the offspring of all creatures arise from their parents, how they are, as Erasmus Darwin said long ago, like separated buds, then we see the truth of the often-made comparison between any or all species of animals and an organism. The individuals of a species are not indeed bound together by protoplasmic strands, but their interdependence is not less complete. A single species utterly destroyed might modify the life of the whole earth. Life itself is dependent upon the invariable presence of minute quantities of iron in the soil. A wandering tribe of savages is an organism not quite so high in the scale of social organisms as is the hydra in the scale of individuals; for the cells of the hydra, although divided broadly into an outer and an inner layer, are yet more divided in their functions than are the members of a savage tribe. For there are only two kinds of person in such a tribe—hunters and cooks; while a highly civilised community, with its immense variety of workmen, is probably not so well organised as any mammal; for there are in such a state thousands of persons, untrained to any special labour, merely a burden to themselves and to the nation.

9. **Origin of Life.**—We have said that life probably began when the conditions of heat and solubility of substance were more favourable to the formation of peculiar and complex matter than at present. But such a statement is often thought to be unphilosophical in view of the fact that we have at present no experience of the formation of such substances, and that it has been conclusively proved that living creatures always proceed from pre-existing life. But those who urge such objections forget that all that has been proved is, that the simplest creatures *known to be alive at present* can be formed only by cell-division one from another, and not from simple chemical materials. But we must remember that those simplest animals are highly developed in comparison with the complex matter from which we conceive life to have sprung; and no one would now expect that such comparatively highly developed animals could arise from simple matter. There is certainly no evidence of the formation at present of the very simplest and original living matter. But, in the first place, could we see it, even with a microscope, if it were to be formed? Might it not be formed molecule by molecule? And, secondly, what chance of survival would such elementary creatures have among the voracious animals that swarm in all places where such simplest creatures might possibly be formed? Instantly they would be devoured, before they could grow large enough to be seen.

Lastly, let it be carefully observed, such a belief as this as to the origin of life, and of the basis of all life in chemical processes, carries with it no necessary adherence to the doctrines of Materialism. The materialist analyses the whole objective world of phenomena into matter and motion. So far, his conclusion is perfectly legitimate; but when he maintains that matter and motion are the only realities of the world, he is making an unwarrantable assumption. Matter in motion is accompanied by consciousness in ourselves. We infer a similar consciousness in creatures like ourselves. As the movements and the matter differ from those that occur within our body, so will the accompanying consciousness. The simplest state of affairs or "body" we can imagine is that of a gas such as hydrogen. But such a simple state of matter may have its accompanying consciousness, as different from ours as is the structure of our bodies from that of a hydrogen molecule. This is, of course, also an assumption, but it is one that harmonises with the facts of experience.

The opposite extreme to Materialism is Idealism, and in this school of philosophy an assumption precisely similar, and exactly opposite to that of Materialism, is made. The idealist says the objective world of phenomena has no existence at all, it is the creation of mind. An objection to such a theory lies in the question, If matter and energy are the creation of mind, how is it that we find them to be indestructible?

Popular philosophy has made an assumption which lies midway between these extremes. It postulates two realities, matter and spirit, having little effect upon one another, but acting harmoniously together.

But the view that is here set forth postulates neither matter nor spirit, but an entity which is known objectively as matter and energy, and subjectively as consciousness. This philosophy goes by the name of Monism. The term consciousness is used for lack of any other to express the constant subjective reality. Carefully speaking, it is, of course, only the more complex subjective processes that form consciousness.

CHAPTER IX

THE DIVIDED LABOURS OF THE BODY

1. *Division of Labour*—2. *The Functions of the Body: Movement; Nutrition; Digestion; Absorption; The Work of the Liver and the Kidneys; Respiration; Circulation; The Changes within the Cells; The Activities of the Nervous System*—3. *Sketch of Psychology*

1. **Division of Labour.**—The simplest animals are one-celled; the higher animals are built up of numberless cells. All the processes of life go on within a single cell. In a many-celled animal the labours of life are divided among the various groups of cells which form tissues and organs. The history of physiological development is the history of this division of labour.

When a dividing cell, instead of separating into two distinct masses, remained, after the division of its nucleus, with the two daughter masses lying side by side, joined together by strands of protoplasm, then the evolution of organic form took a distinct step upwards, and at the same time arose the possibility of greater activity, by means of the division of labour. For when the process had resulted in the formation of an organism of a few dozen cells, arranged very likely in the form of a cup, the outer cells might devote the greater part of their energies to movement and the inner cells to the digestion of food. In the common *Hydra* the body consists of two layers of cells arranged to form a tube, the mouth of which is encircled by tentacles.

The cells of the outer layer are protective, nervous, and muscular; the cells of the inner layer are digestive and muscular. The cells of *Hydra* are therefore not so many-sided in function as are Amœbæ. In animals higher than the simplest worms, a middle layer of cells is always formed which discharges muscular, supporting, and other functions.

With advancing complexity of structure the specialisation of certain cells for the performance of certain functions has become more pronounced. In the human body the division of labour has reached a state of great perfection; we shall give a slight sketch of its arrangements.

2. **The Functions of the Body.**—Our objective life consists of movement, and of feeding to supply the energy for that movement. Growth, reproduction, and decay are elsewhere treated of.

Movement.—We move by the contraction of cells massed into tissues called muscles. Contractility is a property of all living matter; in muscle-cells this function is predominant. This is all that need be said here of movement; the processes of nutrition we must follow more closely.

Nutrition.—All the cells of our bodies are nourished by the stream of fluid food-stuff, the blood, which flows in a number of vessels called arteries, veins, or capillaries, according to their place in the system. From this stream each cell picks out its food; and into another stream—the lymph stream—moving in separate channels—the lymphatics, which, however, join the blood channels, each cell casts its waste material; just as a single-celled animal takes food from the water in which it lives and casts its waste into it.

Nutrition must therefore consist of two series of activities. One series will have for its object the preparation of food-matter so that it may enter the blood, and the excretion of waste products out of the blood. The other series will consist of the activities of the individual cells,—the manner in which they feed themselves.

The first step in the preparation of the blood is digestion. Most food-stuff is solid and indiffusible; before it can enter the blood it must be made soluble and diffusible. The

CHAP. IX *The Divided Labours of the Body* 145

supply of oxygen to the tissues is also a part of these first processes of nutrition. Being a gas, it is treated in a special way which will be described immediately.

Digestion.—The various food-stuffs have various chemical qualities. After being swallowed they enter a long tube, the digestive tract or alimentary canal. Within this canal they are subjected to the action of various digestive juices prepared by masses of cells called glands. Saliva is one of these juices, gastric juice is another, pancreatic juice is another. The effect of these juices upon the food is that most of it is dissolved in the juice and made diffusible. Thus we see an example of the division of labour. An amœba flows round a solid particle of food and digests it. In the higher animals the cells of the digestive glands are specialised for this particular function and do little else.

Absorption.—After the food is digested it leaves the alimentary canal, and is absorbed into the blood-vessels and lymphatics in the walls of the canal. Absorption is not a mere process of diffusion. It is diffusion modified by the cells lining the alimentary tract. Certain chemical changes are effected at the same time. Most of the absorbed food passes to the liver; but the fat does not go directly into the blood, being first absorbed into that other system of vessels, the lymphatics. Eventually it also gets into the blood; for the two streams are connected.

The Work of the Liver and the Kidneys.—The cells of the liver secrete a juice called bile, which is poured into the alimentary canal. The exact function of this juice is still doubtful. It has a certain use in the digestion of fats, but it is largely an excretion. The stream of food-stuff going to the liver contains sugar, the result of the digestion of carbohydrates; albumen, the result of the digestion and absorption of proteids; and certain waste nitrogenous matters formed during the digestion of proteids.

The cells of the liver retain the sugar, store it within themselves, in the same sort of way that a potato stores up starch, and give it up gradually to the blood again. So far as is known they do not affect the albumen in any way, but the waste nitrogenous matter is altered and then sent on in the blood stream to the kidneys.

The cells of the kidneys take this stuff, which was prepared in the liver, and other waste nitrogenous products out of the blood, pass them and a certain amount of water along to the urinary bladder, which empties itself from time to time.

Respiration.—Breathing consists of two distinct acts, inspiration and expiration. During an inspiration air is drawn into lungs. Thence the oxygen passes by diffusion, modified by the fact that the essential membrane is a living one, into the blood. There it enters into a loose combination with hæmoglobin, the red colouring matter of the blood cells, and is thus carried to the cells of the tissues to be absorbed into their living matter. During an expiration we breathe out air which has less oxygen and more carbonic acid gas than normal air. The carbonic acid is a waste-product formed by the cells of the body; it first enters the blood, is then carried to the lungs, and leaves the blood-vessels by a process of diffusion similar to that by which the oxygen entered. The close association of these two processes is simply due to the fact that an organ fitted for the diffusion of one gas in one direction will do for the diffusion of all gases in any direction.

Circulation.—The blood is maintained in a healthy state by the processes we have described. By the active contraction of the heart it is pumped round and round the body, continually carrying with it fresh food to the tissues, and carrying away with it the waste matter cast out of the tissues. All the blood-vessels, except the very smallest, have muscular walls. The heart is a large hollow mass of muscles, is a part of a pair of large blood-vessels that have been bent upon themselves, and arranged so as to form four separate chambers, two upper and two lower, an upper and a lower opening directly into one another on each side. By the contraction of the lower chamber of the left pair the blood is forced through all the vessels of the body; these collect and empty the blood into the upper chamber of the right pair; from this it passes into the lower chamber on the same side, and from this it is forced through the vessels of the lungs, returning to the upper chamber of the left side, and so to the lower chamber of the left side.

The way in which the blood is able to nourish the tissues is as follows:—The outgoing vessels—arteries—enter each mass of tissue; within it they break up into numberless very small, very thin-walled vessels—capillaries; the blood oozes through these into the small spaces—lymph spaces—that occur throughout the tissues; adjacent to these spaces are the cells, which take from the lymph—the fluid that fills the small spaces and the vessels connected with them—what they need, and cast into it their waste. The lymph spaces open into lymph vessels, which, as has been noted, join the blood-vessels. Oxygen and carbonic acid, being gases, pass directly from the blood through the walls of the vessels to the tissues, and from the tissues to the blood.

The Changes within the Cells.—In speaking of protoplasm an outline of the kind of knowledge that we possess of the chemical changes that take place within the cells has been given. We know little more than the substances that enter the cells and the substances that leave them.

Perhaps even this is too much to say; more exactly, we know the substances that enter the body by the mouth and nose, and through the alimentary canal and lungs; we know the substances that leave the body through the kidneys, and, in expiration, through the nose. A large amount of water and traces of other matters leave the body as perspiration; but the chief use of sweating is probably the regulation of the temperature of the body, and the skin should not be thought of as an excreting organ in the same way that the kidneys are. The undigested matter that passes from the food-canal has never been within the blood, and does not therefore concern us in this inquiry.

But we know very little more than this; the analysis of the precise changes that any particular mass of tissue exerts upon the blood—*i.e.* the differences that must exist between the substances entering it and the substances leaving it—is very difficult of determination, because of the immense quantity of blood that passes through any tissue in a short time. This concludes our sketch of the interchange of matter within the body.

The Activities of the Nervous System.—We have now to consider the arrangement of the nervous system—first merely as the means by which all the varied activities of the tissues of the separate parts of the body are co-ordinated and wrought into an harmonious series of actions, and then as the associate of consciousness and of mental processes.

Just as protoplasm may be called the physical basis of life, so is nervous tissue *par excellence* the physical basis of consciousness and of mind. Throughout the whole animal kingdom it has a superficial similarity of structure, and consists of the same three parts.

(1) First there are cells adapted to receive notice of change in the outer world. Changes in the surrounding medium and affecting such cells are called stimuli. These cells sensitive to stimuli form the chief part of the sense organs — the eyes, ears, nose, tongue. Also in the skin there are cells sensitive to alterations of touch and temperature. The effect of stimuli upon such cells is probably to set them into a state of molecular agitation, which may or may not result in chemical changes.

(2) There are connecting fibres or nerves, which, being connected with the sensory cells, take up the vibrations or possibly the chemical changes of the sensory cells and transmit them to the "centre."

(3) There are "central" cells, in which the nerves end, and which are set in molecular agitation by the vibrations of the nerves. This molecular agitation is often, when the central cells are in the brain, accompanied by consciousness. Apparently also agitations may arise "spontaneously" within these central cells and stimulate

the outgoing nerves, and cause muscular movements, or the activity of glands, or other cellular activities.

In many ways analogous to the nervous system is the telegraph system of a country; the receiving stations are the nerve cells, among which are the cells of the sense organs; the connecting wires are the nerve fibres; the central stations are the groups of cells called ganglia, the chief of which are in the brain and spinal cord. The less important ganglia are like the branch offices, they receive messages and transmit them unaltered; the higher ganglia are like the offices of a government, in which messages are received, plans elaborated on the strength of the news, and orders sent out to various parts of the country. All such actions when they take place in the nervous system are called reflex actions, whether a received message be sent on unaltered, or whether the receiving cell regulates the transmission according to the needs of the parts of the body. The analogy of telegraph stations, even with the living clerk to work them and with responsible persons to direct the clerk, does not give a much too complex idea of the activities of nerve cells.

3. **Sketch of Psychology.**—The following is largely derived from Professor Lloyd Morgan's *Animal Life and Intelligence*, to which we refer the reader, and to which we acknowledge our indebtedness.

It very often happens that changes in nervous matter, caused by changes in the outer world, result in what we call a change of consciousness.

Consciousness is the subjective side of molecular disturbance in brain or other nerve matter.

Changes in the outer world which cause disturbance of nerve matter are called stimuli.

Changes of consciousness produced by stimuli are called impressions.

If impressions left no permanent trace in consciousness behind them, there would be no such thing as mind. For mind is based upon memory, and memory is the revival of past impressions, which, we must suppose, have in some way been stored within the cells of the brain in the form, from the objective point of view, of a certain arrangement of its particles.

When, after the revival of past impressions, we are able to discriminate between them, we call them sensations.

Now sensations are referred, in consciousness, not to

FIG. 31.—Attitude of a hen protecting her brood against a dog
(From Darwin's *Expression of Emotions*.)

the brain cells which discriminate between them, but to the cells of the sense organs which received them.

Further, we refer, by experience, the causes of sensations to the outer world. We do this by a mental process which is called perception.

Now out of perceptions, and through associations, there arise expectations. The mental process involved in the formation of an expectation is called inference.

Inference is of two kinds :
(1) Perceptual, drawn from direct experience, as in the inference as to a rain storm from a black cloud.
(2) Conceptual, which, though based upon experience, yet can predict events that have never been experienced. For instance, one who had studied in books only the causes of volcanic activity, might predict with a certain amount of confidence a flow of lava from a volcano, when he saw it in that state of activity which he knew usually preceded an eruption of lava.

What we call the emotions, love, hate, fear, and others, are, so far as we can tell, agitations of nervous matter which affect consciousness. Their exciting stimuli—inferences for example—proceed, immediately, from within the brain, ultimately, from changes in the outer world.

We have, therefore, the following orders of consciousness, which are easily distinguishable in theory :
(1) Impressions, or the effects of environmental changes upon nervous matter; the retaining and revival of these constitutes the basis of memory.
(2) Sensations, which occur when the differences that exist between impressions are discriminated.
(3) Perceptions, which are the outward projections into the world, by mental acts, of the molecular disturbances caused in the brain by environmental changes. For example, light falls upon the retina, stimulates the optic nerve, and causes a molecular disturbance in the brain, but the consciousness excited in us is not of the brain disturbance, but of the light.

It is most essential that the distinction between perceptual and conceptual inference be clearly realised, as it is probable that it is the faculty for the latter which more than anything else separates man from the lower animals. We may be nearly certain that many animals exercise perceptual inference, and we may affirm with little doubt that none has ever performed a conceptual one. It has been stated that a monkey that had learned to screw and unscrew a handle from a broom had learned "the principle of the screw." This is entirely erroneous. The monkey merely observed that a certain movement given to the handle caused it to separate from the broom, and inferred perceptually that the same result would always follow from the same action.

It is evident that a sound comparative psychology of the animal kingdom, or even of a few of the highest species, is beyond the present possibilities of science.

CHAPTER X

INSTINCT

1. General Usage of the Term—2. Careful Usage of the Term—3. Examples of Instinct—4. The Origin of Instinct

IN considering the mental life of animals, we must settle how far it is comparable to that of man. We judge of the mental processes of human beings, other than ourselves, from their actions; and we can only do the same when dealing with animals. If we often err in inferring the mental states of our fellow-men, how much more are we liable to error when we are considering creatures different from ourselves. Still, believing as we do in the continuity of life, both objective and subjective, by careful proceeding it is probable that in time we may arrive at a certain state of precision and exactness in comparative psychology.

Since the idea that is formed of the world is gained entirely from sensations, the world of every creature must be largely constructed from its dominant sense; in a dog, for instance, from scent.

In common speech the actions of animals are all ascribed to instinct. The notion which underlies the term is, that while the actions of men are determined by reason, those of animals are prompted by a blind power of doing that which is fitted to the successful conduct of their lives. This, as we shall see, is a notion that requires modification.

1. **General Usage of the term Instinct.**—Every one has a general notion of what is meant by instinct, but few are

agreed as to the precise usage of the word; thus when the birds build their nests, or when the bees collect honey and form their combs, their acts are with one accord said to be instinctive; but some would demur at using such a term to describe the love of parents for their children, the courage of brave men, or the artist's perception of beauty. But, even supposing we agree to mean by instinct all those actions which are neither simply reflex nor purely rational, there will still remain great difference of opinion as to its origin. Thus the love of parents will not be imagined as due to practice, either in the individual or its ancestors, but rather to take origin in some hidden necessity of nature; while the rapid closure of the eyes as protection from an expected blow would seem in all likelihood to have begun in a rapid exercise of intelligence, which, by being often repeated, had ceased to be accompanied by conscious effort.

It seems to us that there is still need of a vast amount of observation and experiment before a theory of the origin of instinct that will be at all satisfactory can be framed. As already remarked, it is not easy to decide even in what sense the term ought to be used. This being so, we shall content ourselves with mapping the field of thought and indicating the lines of inquiry that must be followed before a just view of the subject will be possible.

If we arrange examples of all the movements of animals in the order in which they are performed in the lifetime of the individuals, not limiting ourselves to those acts which involve the whole organism, but considering also those which a single organ or mass of tissue may execute, we shall see at a glance all the possible varieties of activity with which we can be concerned. It is, of course, only the movements of comparatively large masses of tissue with which we can deal at present. The molecular movements which lie at the base of all the visible ones are as yet almost unknown.

Even before birth, visible movements of the parts of the higher animals occur; as, for instance, the beating of the heart. Such movements may be either "automatic" or reflex. At birth, in addition to such movements of its parts, the organism acts as a whole; it reacts to its environment, and in time performs "voluntary" actions.

The acts of the parts of an organism may be—
  (1) "Automatic," as, for example, the beating of the heart.
  (2) Reflex, as, for example, the intestinal movements which force the food through the alimentary canal, or the movements involved in sneezing.
  (3) Mixed actions which are partly automatic and partly reflex, such as the respiratory movements.

The movements of the entire organism may be of a very complex nature. They may be—
  (1) Reflex; as when we start at a sudden noise.
  (2) "Innate," commonly called instinctive; these are best observed in newly-born animals, for in them intelligence, which must be based upon experience, is necessarily at a minimum.
  (3) "Habitual," such as are rapidly learned and are then performed without mental effort, which imply an innate capacity, and are therefore allied to (2).
  (4) Intelligent, such as imply mental activity, which consists in the combining and rearranging all the other possible acts of the order—(1), (2), or (3); and which may be recognised in all adaptations to novel circumstances.

This classification possesses most obvious faults, but it has certain advantages. It reveals some of the difficulties that delay the would-be definer of instinct. For the essential criterion of an instinctive action is that all the machinery for its performance, as a reflex to a certain stimulus, lies ready formed within the organism; but the apparently insoluble questions present themselves, How soon may not actions be modified by intelligence? and How in a mature animal with considerable experience is one to separate the purely instinctive acts from the intelligently modified instinctive acts?

Also it is evident that "habitual" actions may be "instinctive" actions deferred until the creature be further developed, as the flight of many birds is deferred; or they may be actions in the formation of which intelligence has had a considerable share.

Now all these activities of an entire organism may be studied from four points of view:—

(1) Of natural history, or general description, such as occurs here and there throughout this work:
(2) Of physiology, or the analysis of the muscular, nervous, and other mechanisms involved; as treated generally in the last chapter:
(3) Of psychology, or the investigation into the states of consciousness and mental processes concerned; as sketched in the last chapter:
(4) Of ætiology, or study of the factors in their origin and development.

We shall first define more carefully than we have yet done what we shall speak of as instinct, then give a few examples, and finally discuss the ætiology of it.

2. **The Careful Usage of the term Instinct.**—We have enumerated all the possible varieties of action, and the possible states of consciousness with which they may be associated have been described in the last chapter. If we retain the use of the term instinct we must explain to what order of activity we shall apply it. In our use of the term we shall not strive after any great precision; for, as already noted, the difficulty of precision seems to us to be at present insurmountable. In a general way we shall call any action which does not require for its execution any immediate exertion of perceptual inference an instinctive action. Thus a burned child dreads the fire; such dread and its consequent avoidance of fires may, with propriety, be termed instinctive. After the first burn the avoidance will, for a short time, be the result of perceptual inference; but in perhaps a few days only the avoidance becomes "instinctive"; or it might be called "habitual," as hinted previously. It is, of course, to be understood that an "instinctive" action is not necessarily the result of this "lapsed intelligence," as it has been called. Thus, when a worm wriggles away from a fire it probably has not at any time reasoned out to itself the advantages of such procedure,

yet it may well be said to avoid the fire instinctively. It is obvious that, if we agree to use the term as defined, we must call all the actions of the lower animals, whose consciousness has never risen to the level of perceptual inference, instinctive. This definition is based upon the assumption that we can determine the conscious states of animals; but, as we have repeatedly said, it is only within very wide limits that we can with any certainty do this. The intention, however, is to preclude all those actions which are certainly or probably rational, and at the same time include adaptive reflex actions.

Mr. Herbert Spencer has defined instinct as compound reflex action, while Mr. Romanes separates it from reflex action and from reason as follows :—

"Reflex action is non-mental neuro-muscular adjustment due to the inherited mechanism of the nervous system, which is found to respond to particular and often-recurring stimuli, by giving rise to particular movements of an adaptive though not of an intentional kind."

"Instinct is reflex action into which there is imported the element of consciousness. The term is therefore a generic one, comprising all those faculties of mind which are concerned in consciousness and adaptive action, antecedent to individual experience, without necessary knowledge of the relation between means employed and ends attained, but similarly performed under similar and frequently-recurring circumstances by all the individuals of the same species."

"Reason or intelligence is the faculty which is concerned in the intentional adaptation of means to ends. It therefore implies the conscious knowledge of the relation between means employed and ends attained, and may be exercised in adaptation to circumstances novel alike to the experience of the individual and that of the species."

Mr. Romanes therefore separates reflex action from instinctive action by limiting the term instinct to those actions which are, as a matter of fact, conscious reflexes. His definition is open to objection on the same ground that ours is, only in a greater degree; for it is easier to determine the presence of perceptual inference than the absence of consciousness; this criterion may be of theoretical interest,—it is of no practical use. The other attributes he enumerates should be carefully studied.

Prof. Lloyd Morgan also separates, but by no hard-and-fast line, the automatic and reflex actions, which are reactions to definite stimuli, from instinctive actions, which, according to him, are " sequences of co-ordinated activities, performed by the individual in common with all the members of the same more or less restricted group, in adaptation to certain circumstances, oft recurring or essential to the continuance of the species."

He separates these from intelligent actions, which are "performed in special adaptation to special circumstances."

Instinctive activities he conceives to be performed "without learning or practice." If the actions need a little practice he calls them "incomplete instincts"; if a great deal of practice be necessary they are called "habitual activities"; if they are not perfectly developed at birth but after further development can be performed without practice they may be called "deferred instincts." A further useful classification of instincts is into "perfect" and "imperfect," according to the precision of their adaptation to the desired end.

Mr. Lloyd Morgan's definition, like the others, implies that one can separate rational from non-rational actions; but he safeguards himself by defining instincts as "oft-recurring or essential to the continuance of the species," in contradistinction to intelligent actions which are performed in special adaptation to special circumstances. It is important to notice that the terms of the definition are that instincts are either oft-recurring or essential, and not oft-recurring and essential, for many instincts are only either one or the other and not both. But it is not always possible to say of a certain action that it is a special adaptation to a special circumstance, and is therefore rational, and not in reality an instinctive adaptation to circumstances that are of frequent occurrence although we have not observed them to be so.

This definition, however, emphasises the fact that instincts are common to species; it is, however, not easy to estimate the exact significance of this fact, for the apparent similarity in the actions of individuals of the same species must, to a certain extent, be due to incompleteness of observation.

It is after considering all these definitions that we have come to the conclusion that it is convenient to describe all those actions of animals which are not immediately rational or intelligent as instincts. If we classify an instinct as reflex in cases where the exact chain of internal events is known and use the other qualifications already enumerated, we reach a simplicity and precision of speech that is convenient.

At the same time all such criteria as adaptiveness and similarity of performance in all the individuals of a species can obviously be applied as they are discovered.

The essential distinction, we believe, between non-intelligent, that is, instinctive, and intelligent actions, is that non-intelligent actions are performed in virtue of the innate and co-ordinated mechanisms of the organism, whereas for an intelligent action the organism has to do a greater or less amount of the co-ordination for itself.

3. **Examples of Instinct.**—If we classify all the actions of animals according to the period of life at which they are performed, we shall find that there are three distinct classes of action which may with convenience be considered separately.

They are—
- (1) Those which are performed at birth, or shortly afterwards, as perfectly, or nearly so, as at any future time;
- (2) Those more varied actions which are characteristic of the mature life of any animal;
- (3) Those which are associated with reproduction.

The first of these classes must evidently consist of very pure instincts, since the creature cannot be supposed to reason before it has any store of experience.

The second class is well typified by the marvellous actions of insects, such as ants, bees, and wasps. These *may* be instinctive, but it is very probable that many of them are, at least, improved by intelligence.

(*a*) They may be perfected by perceptual inferences on the part of the individuals, and the mental efforts may or may not, after a certain number of repetitions, be replaced by reflexes.

(*b*) They may be perfected by less complex mental efforts, such as those involved in imitation or in receiving instruction from other members of the species.

Actions of the third class may be as purely instinctive as any of those in the first class, or may be improved by intelligence like those of the second class; but among them are many of the most wonderful performances of animals, for they often seem to show a prevision of an unknown future.

Some interesting experiments have been made upon instincts of the first class. The observations show that the precision of the neuro-muscular co-ordinations of some newly-born creatures is very surprising. Mr. Spalding blindfolded some chickens immediately after they were hatched, and removed the hood after two or three days when they were stronger. He says that " almost invariably they seemed a little stunned by the light, remained motionless for several minutes, and continued for some time less active than before they were unhooded. Their behaviour, however, was in every case conclusive against the theory that the perceptions of distance and direction by the eye are the result of experience, or of associations formed in the history of each individual life."

" Often at the end of two minutes they followed with their eyes the movements of crawling insects, turning their heads with all the precision of an old fowl. In from two to fifteen minutes they pecked at some speck or insect, showing not merely an instinctive perception of distance, but an original ability to judge, to measure distance, with something like infallible accuracy. They did not attempt to seize things beyond their reach, as babies are said to grasp at the moon, and they may be said to have invariably hit the objects at which they struck, they never missed by a hair's-breadth, and that too when the specks at which they aimed were no bigger and less visible than the smallest dot of an *i*. To seize between the points of the mandibles at the very instant of striking, seemed a more difficult operation. I have seen a chicken seize and swallow an insect at the first attempt ; most frequently, however, they struck five or six times, lifting once or twice before they succeeded in swallowing their first food." Again, " The art of scraping in search of food, which, if anything, might be acquired by imitation, for a hen with chickens spends the half of her time in scratching for them, is nevertheless another indisputable case of instinct. Without any opportunities of imitation, when kept quite isolated from their kind, chickens began to scrape when from two to six days old. Generally the condition of the ground was suggestive ; but I have several times seen the first attempt, which consists of a sort of nervous dance, made on a smooth table." Another experimenter " hatched out some chickens on a carpet, where he kept them for several days. They showed no inclination to scrape, because the stimulus supplied by the carpet to the soles of their feet was of too novel a character to call into action the hereditary instinct ; but when a little gravel was sprinkled on the carpet, and so the appropriate or customary stimulus supplied, the chickens immediately began their scraping movements."

Another instance of the first class of instincts is the fear said to be shown by many animals for their natural foes ; but on this point we find a certain conflict of evidence. Thus kittens are said to show disgust at a dog, and, while still blind, at a hand that has touched and smells of a dog, or to tremble with excitement at the smell of a mouse. A chicken or young turkey will show evident signs of fear at hearing the cry of a hawk. Ants of various species that are mutually hostile recognise an enemy, and fight ; but, on the other hand, there are observations to the effect that, if taken young enough, ants of several such species may be brought up together as a happy family.

The instinctive tameness or wildness of many animals towards man is probably the effect of intelligence and information given to one another ; as is the avoidance of the same kind of trap, after a short experience of its properties, by

M

all the mice of a house or birds of a district. The wild rabbit is extremely timid, but the domesticated variety is as tame as possible. In explanation of such cases we might easily invoke the aid of "the principle of cessation of natural selection" when the safety of the species ceased to depend upon wildness, but we prefer to suppose that the direct action of intelligence is to a great extent operative.

As already noted, what are perhaps the most striking examples of instinct of the second class occur among insects. The comb-building of bees, the wars, the slave-using, the agricultural pursuits of ants, have been so often described that they need not detain us here. The brain of an ant was to Darwin the most wonderful piece of matter in the world. Wonderful, indeed, it would be if we supposed that all the acts of an ant were truly instinctive, that is, that the nervous machinery of co-ordination was ready, waiting only the appropriate stimulus to evoke any one of that series of nicely-adjusted actions. But if we suppose that individual intelligence has a considerable share in that co-ordination, then the brain of an ant, though still very wonderful, is not to us quite so astounding an arrangement of particles as it was to Darwin.

The third class of instincts, those connected with reproduction, comprise such actions as the building of homes and nests, the storage of food for the use of young that may never be seen, and the care of young after birth.

The nest-building of birds would form a very good subject upon which to experiment, in order to determine how far such a complex act may be truly instinctive, and how far it is perfected by training, by imitation, and by intelligent practice and observation. The method would be to isolate young birds from their parents and from all other creatures of their kind, so as to preclude training and imitation, and then see how far the nests that they built resembled the typical nest of their species. Then one might remove other birds from their parents but allow them the society of the members of an allied species, but one whose nests differed to some observable extent from those of the species under experiment. So far as we know, no adequate experiments have been carried out.

It is stated that the nests of British birds let loose in Australia differ very greatly from the nests that they would build at home. Now this may be due to the absence of training and the possibilities of imitating the specific nest, or it may be due to the absence of the materials with which to build the characteristic nest.

When we consider the careful provision that the *Sphex* wasp makes for its young, young which she will never see ; or the selection by a butterfly of the leaf upon which she lays her eggs, the only sort of leaf, it may be, that will serve the grubs, when hatched, for food, food of a kind which the butterfly herself does not eat, we have before us examples of instincts most wonderful in their perfection and most obscure in their origin. The ideas involved are too complex for us to believe that the actions can be the result of intelligence. So far as we can see at present, such instincts can only be accounted for by the Natural Selection of fortunate varieties of habit. The care of young and the habit of incubation are instincts upon which a certain amount of thought has been bestowed. For ourselves, we incline to the idea which may seem mystical to some, that such habits are born of an inevitable affection for what is so nearly related to the very body of the parents. To escape an explanation of such habits in terms of affection, certain naturalists have suggested the demonstrably absurd notion that birds sit upon their eggs in order to cool a fevered breast. If such were her object, the very worst place that a bird could select as her seat would be a woolly hairy nest containing eggs which, in virtue of internal chemical changes, generate heat of themselves.

4. **The Origin of Instinct.**—The theory of the evolution of instinct has been worked at by Darwin, and since his day by various other authors, notably Mr. Romanes, Mr. A. R. Wallace, and Professor Eimer ; while Professor Weismann's doctrines have necessitated the revision of certain plausible hypotheses.

Darwin, of course, supposed that Natural Selection

was the means of the evolution of habit as much as of form.

Mr. Romanes, starting from this as a basis, has constructed a well-reasoned and lucid theory. He supposes that while many instincts have been evolved by Natural Selection, such instincts being called Primary, other habits become instinctive through the "lapse of intelligence." Actions performed at first with mental effort, becoming after sufficient repetition so ingrained upon the nervous system that a mechanism of neuro-muscular co-ordination has been established, are referred to as Secondary Instincts. He imagines also a third class of Mixed Instincts in which there are primary instincts that have been altered and improved by intelligent variations of habit, or secondary instincts that have been modified by natural selection.

Obviously, therefore, he supposes that intelligence may be a factor in the formation of any habit that may be under consideration.

But this theory of instinct becomes impossible if we accept Professor Weismann's doctrine that acquired characters can not be inherited (this doctrine will be discussed in a later chapter). If this be true, the only possibilities are primary instincts, and secondary instincts formed afresh during each individual lifetime, and mixed instincts of the same nature.

The exact antithesis to Professor Weismann's theory is upheld by Professor Eimer, who believes that instincts have been evolved chiefly by the perpetuation of what Mr. Romanes calls secondary instincts. There is little evidence that this is the case. The value of Eimer's work really lies in his insistence upon the intelligent action of animals as apart from purely instinctive action.

Mr. Wallace has begun the analysis of the particular forms which the intelligence of animals takes. He supposes that imitation of parents and other members of the species has a great influence upon the actions of individuals. He has dwelt especially upon such cases as the song and nest-building of birds.

It may be pointed out as a matter for consideration that, granted that parents teach their offspring, as, for instance,

CHAP. X   *Instinct*   165

birds teach their fledglings to fly, and ants their young their place in the community of the nest, and that animals imitate each other, it is quite possible, and indeed probable, that an instinct may be steadily improved throughout suc-

FIG. 32.—Young ducks catching moths. (From St. John's *Wild Sports*.)

cessive generations by the intelligence of the individuals of a species, without any acquired character being inherited.

The possible factors in the evolution of instinct are therefore—

(1) Natural Selection, which might develop innate capacity; this is certainly insufficient for the development of form, and therefore, probably, also of mind.

(2) Individual intelligence, which directly modifies the actions of individuals, and is also used when, by imitation and education, the habits of a species are steadily improved.

(3) The "lapsing of intelligence," forming "secondary," and helping to form "mixed" instincts.

But the probable factors are the first two.

PART III

THE FORMS OF ANIMAL LIFE

CHAPTER XI

THE ELEMENTS OF STRUCTURE

1. *The Resemblances* and *Contrasts between Plants and Animals*—2. *The Relation of the simplest Animals to those which are more complex*—3. *The Parts of the Animal Body: Organs, Tissues, Cells*

THE study of form and structure (Morphology), and the study of habit and function (Physiology), are both as essential to science as the realities are in life. It is with the forms of animal life and their structure that we have now to do, but it seems useful at the outset to compare plants and animals.

1. **The Resemblances and Contrasts between Plants and Animals.**—Every one could point out some differences between a tree and a horse, but many might be puzzled to distinguish clearly between a sponge and a mushroom, while all have to confess their inability to draw a firm line between the simplest plants and the simplest animals. For the tree of life is double like the letter V, with divergent branches, the ends of which, represented, let us say, by a daisy and a bird, are far apart, while the bases gradually approach and unite in a common root.

Plants and animals are alike, though not equally, alive.

Diverse as are the styles of animal and vegetable architecture, the materials are virtually the same, and the individuals in both cases grow from equally simple beginnings.

Even movement, the chief characteristic of animals, occurs commonly, though in a less degree, among plants. Young shoots move round in leisurely circles, twining stems and tendrils bend and bow as they climb, leaves rise and sink, flowers open and close with the growing and waning light of day. Tendrils twine round the lightest threads, the leaves of the sensitive plant respond to a gentle touch, the tentacles of the sundew and the hairs of the fly-trap compare well with the sensitive structures of many animals. The stamens of not a few flowers move when jostled by the legs of insects, and the stigma of the musk closes on the pollen.

Plants and animals alike consist of cells or unit masses of living matter. The structure of the cell and the apparent structure of the living stuff is much the same in both. We may liken plants and animals to two analogous manufactories, both very complex; we study the raw materials which pass in, many of the stages and by-products of manufacture, and the waste which is laid aside or thrown out, but in neither case can we enter the secret room where the mystery of the process is hidden.

In the pond we find the eggs of water-snails and water-insects attached to the floating leaves of plants; in the ditches in spring we see in many places the abundant spawn of frogs and toads; we are familiar with the heavily yolk-laden eggs of birds. Now, with a little care it is quite easy to convince ourselves that an egg or ovum is to begin with a simple mass of matter, in part, at least, alive, and that by division after division the egg gives rise to a young animal. We are also well aware that in most cases the egg-cell, for cell it is, only begins to divide after it has been penetrated, and in some subtle way stimulated, by a male unit or sperm. The great facts of individual history or development then are, the apparent simplicity in the beginning, the preliminary condition that the egg-cell be united with a male unit, and the mode of growth by repeated division of the ovum and its daughter-cells. In those plants with which we are most familiar, the facts seem different, for we watch bean and oak growing from seeds which, instead of being simple units, are very complex structures. But the seed is not the beginning of a plant, it has already a long history behind it, and when that history is traced back to the seed-box and possible seeds of the parent plant, there it will be seen that the beginning of the future herb or tree is a single cell. This is the equivalent of the animal ovum, and, like it, begins its course of repeated divisions after it has been joined by a kernel or nucleus from the pollen grain.

Thus, to sum up, along three different paths we reach the same conclusion, that there is a fundamental unity between plants and animals. In the essential activities of their life, in the stones and mortar of their structure, and lastly, in the way in which each individual begins and grows, there is a real unity.

Yet, after all, plants and animals *are* very different. The two kinds of organisms may be ranked as two great branches of one tree of life, yet the branches diverge widely and bear different foliage. The facts of divergence and diversity are as undeniable as the inseparable unity of the basal trunk and the genuine sameness of life throughout the whole tree. I have stated the chief contrasts between plants and animals in a tabulated summary—

Some Exceptions	Characteristics of Animals	Characteristics of Plants	
Some Protozoa and parasites simply absorb	They feed on more or less solid food	They absorb soluble food	
Some green Protozoa (etc.) seem to be able to utilise carbonic acid as plants do	They obtain the requisite carbon from starch, sugar, fat, etc., made by plants or by other animals	They obtain the requisite carbon from carbonic acid gas in the air or water	Car... an... ot... su...
Again, some Protozoa are probably able to feed like plants	They obtain the requisite nitrogen from nitrogenous compounds not simpler than albuminoids, made by other organisms. Most of them are known to get rid of nitrogenous waste-products	They obtain the requisite nitrogen from simple nitrogenous compounds, especially the nitrates of the soil. They do not get rid of nitrogenous waste-products	Aga... fun... are... the...
A few (e.g. some Protozoa, the freshwater sponge, the *Hydra*, etc.) have green pigment very closely analogous to or identical with chlorophyll-green	They have very rarely any chlorophyll	The majority possess chlorophyll, the green pigment by aid of which the living matter utilises the energy of sunlight in reducing carbonic acid (with liberation of oxygen) and in building up complex substances	Fun... ha...
Cellulose seems to occur in some Infusorians, and forms most of the tunic or cuticle of the passive sea-squirts or Ascidians	They consist of cells which often have no very definite cell-walls, rarely have them demonstrably different from the cell-substance, and almost never show any trace of cellulose	The component cells are walled in by cellulose, a material chemically allied to starch	Som... a t...
	Marked division of labour among the cells is characteristic	The cells exhibit on an average much less division of labour	
	They utilise food-material already worked up by plants or by other animals; they convert this potential energy into kinetic energy in locomotion and external work; they are characteristically oxidisers, and are predominantly active	They build up crude, chemically simple food-material into complex substances; they convert the kinetic energy of sunlight into the potential chemical energy of these complex food-stuffs; they are characteristically reducers (of carbonic acid), expend comparatively little energy in motion or external work, and are predominantly passive	

The net result of this contrast is that animals are more active than plants. Life slumbers in the plant; it wakes and works in the animal. The changes associated with the living matter of an animal are seemingly more intense and rapid; the ratio of disruptive, power-expending changes to constructive power-accumulating changes is greater; most animals live more nearly up to their income than most plants do. They live on richer food; they take the pounds which plants have accumulated in pence, and spend them. Of course plants also expend energy, but for the most part within their own bodies; they neither toil nor spin. They stoop to conquer the elements of the inorganic world, but have comparatively little power of moving or feeling. They are more conservative and miserly than the liberally spendthrift animals, and it is possible that some of the most characteristic possessions of plants, *e.g.* cellulose, may be chemical expressions of a marked preponderance of constructive and up-building vital processes. It is enough, however, if we have to some extent realised the commonplaces that plants and animals live the same sort of life, but that the animals are on an average more active and wide-awake than the plants.

2. **The Relation of the Simplest Animals to those which are more Complex.**—From the pond-water catch in a glass tube one of the small animals, suppose it be a tiny water-flea or a minute " worm "; how does it differ from one of the simplest animals, such as an Infusorian? It consists of many units of living matter instead of only one. The contrast is like that between an egg and the bird which is hatched from within it. The simplest animals are single cells, all the others from sponge to man are many-celled. The Protozoa are units; all others—the Metazoa—are composite aggregates of units, or cities of cells.

Compare the life of one of the Protozoa with that of a worm, a frog, or a bird. Both are alive, both may be seen moving, shrinking away from what is hurtful, drawing near to what is useful, engulfing food, and getting rid of refuse. Both are breathing, for carbonic acid will poison them, and dearth of oxygen will kill them; both grow and multiply. But in the single-celled Protozoon all the processes of life occur within a unit mass of living matter. In the many-celled Metazoon the various processes occur at different parts of the body, are discharged by special sets of cells, among which the labour of life has been divided. The life of the Protozoon is like that of a one-roomed house which is at once kitchen and work-room, nursery and coal-cellar. The life of the Metazoon is like that of a mansion where there are special rooms for diverse purposes.

In having no "body" the Protozoa are to some extent relieved from the necessity of death. Within the compass of a single cell they perform a crowd of functions, but tear and wear are often made good again, the units have great power of self-recuperation. They may, indeed, be crushed to powder, and they lead no charmed life safe from the appetite of higher forms. But these are violent deaths. What Weismann and others have insisted on is that the unicellular Protozoa, in natural conditions, need never die a natural death, being in that sense immortal. It is true that a Protozoon may multiply by dividing into two or more parts, but only a sort of metaphysical individuality is thus lost, and there is nothing left to bury. We would not, however, give much prominence to a strange idea of this kind. For the "immortality of the Protozoa" is little more than a verbal quibble; it amounts to saying that our common idea of death, as a change which makes a living body a corpse, is hardly applicable to the unit organisms. I believe, moreover, that the idea has been exaggerated; for instance, the Protozoa in the open sea, in their natural conditions, seem to die in large numbers.

The combination of all the vital activities within the compass of a single-cell involves a very complex life within the unit,—not more complex than the entire life of a many-celled animal, but fuller than that of one of its component cells. While a Protozoon is relatively simple in structure, its life of crowded functions, such as moving, digesting, breathing, is exceedingly complex. The simpler an organism is in structure the more difficult will it be to study its separate functions. Physiological or functional simplicity is in inverse

The Elements of Structure

ratio to structural or morphological simplicity. Thus the physiologist makes most progress when he seeks to understand animals with many parts, for there he can find a large number of units, all as it were working at one task. The life of a Protozoon is more manifold and complex than that of any unit from a higher animal, just as the daily life of the savage—at once hunter, shepherd, warrior—is more varied than ours.

Already it has been recognised that every many-celled animal begins its life as a single cell,—as an egg-cell with which a male element has united. Every Metazoon begins its life as a Protozoon, no matter how large the animal, for the whales arise from ova "no larger than fern-seed," no matter how lofty the result, for man himself has to begin his life at the literal beginning. The fertilised egg-cell divides and re-divides, its daughter-cells also divide, the resultant units are arranged in layers, clubbed together to form tissues, compacted to form young organs, and the result is such a multicellular body as we possess; but while this body-making proceeds, certain units are kept apart, in some way insulated from the process of growth, to form the future reproductive elements, which, freed from the adult body, will begin a new generation. Back to the beginning again every Metazoon has to go, and if we believe that the Protozoa are not only the simplest, but also represent the first animals, we have here the first and perhaps most important illustration of the fact that in its development the individual more or less recapitulates the history of the race. The simplest animals are directly comparable with the reproductive cells of higher animals, but the divided cells of the ovum remain clubbed together to form a young animal, while the daughter-cells of a Protozoon separate from one another, each as a new life.

The gulf between the single-celled and many-celled animals is a deep one, but it has been bridged. Otherwise we should not exist. Traces of the bridge now remain in what are called "colonial Protozoa," which, however troublesome to those who like crisp distinctions, are most instructive to those who would appreciate the continuity of the tree of life. These exceptional Protozoa are loose colonies of cells, descendants or daughter-cells of a parent unit, which have remained persistently associated instead of going free with the usual individualism of Protozoa. They illustrate to some minds a primitive co-operation of cells; they show us how the Metazoa or multicellular animals may have arisen.

3. **The Parts of the Animal Body.**—The physiologist investigates life or activity at different levels, passing from his study of the animal as a unity with habits and a temperament, to consider it as an engine of organs, a web of tissues, a city of cells, or finally as a whirlpool of living matter. So the morphologist investigates the form of the intact animal, then in succession its organs, their component tissues, the minuter elements or cells, and finally the structure of the living stuff itself. Moreover, as there is no real difference between studying a corpse and a fossil, the palæontologist is also among the students of morphology; and most of embryology consists of studies of structure at different stages in the animal's life-history.

The outer *form* of normal animals seems to be always artistically harmonious. It has a certain hardly definable crystalline perfection which pleases our eyes, but those who have not already perceived this will not see much meaning in the assertion, nor in Samuel Butler's opinion that "form is mind made manifest in flesh through action."

"I believe a leaf of grass is no less than the journey-work of the stars,
And the pismire is equally perfect, and the grain of sand, and the egg of the wren,
And the tree-toad is a chef-d'œuvre for the highest,
And the running blackberry would adorn the parlours of heaven,
And the narrowest hinge in my hand puts to scorn all machinery,
And the cow crunching with depressed head surpasses any statue,
And a mouse is miracle enough to stagger sextillions of infidels!"
                                            WALT WHITMAN.

It is also important to think of the different kinds of symmetry, how for instance the radiating sea-anemones and jellyfishes, which are the same all round, differ markedly from bilaterally symmetrical worms, lobsters, fishes, and most other animals. Then there is the difference between unsegmented animals which are all one piece (like the lower worms and the molluscs), and those whose bodies consist, as in earthworm and crayfish, of a series of more or less similar rings or segments, due to conditions of growth of which we know almost nothing.

Organs are well-defined parts, such as limb or liver, heart or brain, in which there is a predominance of one or a few kinds of vital activity. Gradually, alike in the individual and in the race, do they take form and function. There is contractility before there are definite contractile organs or muscles; there is diffuse sensitiveness before there are defined nerves or sense-organs. The progress of structure, alike in the individual and in the race, is from simplicity to complexity, as the progress of function is from homogeneous diffuseness to heterogeneous specialisation. The two great kinds of progress may be illustrated by contrasting a sea-anemone and a bird. The higher animal has more numerous parts or organs, the division of labour within its body has brought about more differentiation of structure, but it is also a more perfect unity, its parts are more thoroughly knit together and harmonised. There is progress in integration as well as in differentiation.

"The shoulder-girdle of the skate," W. K. Parker says, "may be compared to a clay model in its first stages, or to the heavy oaken furniture of our forefathers that stood ponderous and fixed by its own massy weight. As we ascend the vertebrate scale, the mass becomes more elegant, more subdivided, and more metamorphosed, until, in the bird class and among mammals, these parts form the framework of limbs than which nothing can be imagined more agile or more apt. So also as regards the sternum; at first a mere outcrop of the feebly developed costal arches in the amphibia, it becomes the keystone of perfect arches in the true reptiles, then the fulcrum of exquisitely constructed organs of flight in the bird; and lastly, forms the mobile front wall of the heaving chest of the highest vertebrate."

Of the *order in which organs appear* or have appeared we can say little. The simplest sponges and polypes are little more than two-layered cups of cells, the cavity of the cup being the primitive food-canal. A parallel stage occurs in the early life-history of most animals, when the embryo has the form of a two-layered sac of cells, or is in technical language a gastrula. Both in the racial and individual life-history the formation of this primitive food-canal occurs very early. But it is not certain that it—the primitive stomach—was not at a still earlier stage an internal brood-cavity!

But instead of speculating about this, let us seek to understand what is meant by the *correlation of organs*. Certain parts of the body stand or fall together, they are physiologically knit, they have been evolved in company. Thus heart and lungs, muscles and nerves, are closely correlated. Sometimes it is obvious why two or three structures should be thus connected, for it is of the very essence of an organism that its parts are members one of another. In other cases the reason of the connection is obscure.

When organs either in the same or in different animals have a similar origin, and are built up on the same fundamental plan, they are called *homologous*. Those whose resemblance is merely that they have similar functions are termed *analogous*. Even Aristotle recognised that some structures apparently different were fundamentally the same, and no small part of the progress of morphology has consisted in the recognition of homologies. Thus it was a great step when Goethe and others showed that the sepals, petals, stamens, and carpels of a flower were really modified leaves, or when Savigny discerned that the three pairs of jaws beside an insect's mouth were really modified legs. To Owen the precision of our conceptions in regard to homologies is in great part due, though subsequent studies in development have added welcome corroboration to many of the comparisons which formerly were based solely on the results of anatomy. Thus an organ derived from the outer embryonic layer cannot be homologous with one derived from the innermost stratum of embryonic cells. Homologous organs in one animal are well illustrated by the nineteen pairs of appendages borne by a crayfish or lobster. These differ greatly in form and in function; many of them are not analogous with their neighbours, one feels and another bites, one seizes and another swims, but they are all homologous. So are the different forms of fore-limb, the pectoral fin of a fish, the fore-leg of a frog or lizard, the wing of a bird, the flipper of a whale, the fore-leg of a tiger, the arm of man. But the wing of an insect is merely analogous not homologous with that of a bird, while the wings of bats and birds are both analogous and homologous.

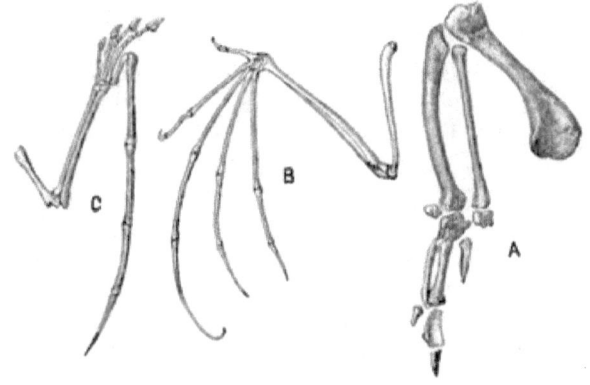

FIG. 33.—Bones of the wing in pigeon (A), bat (B), extinct pterodactyl (C). (From Chambers's *Encyclop.*)

Change of Function.—Organs are not mechanisms rigidly adapted for only one purpose. In most cases they have a main function and several subsidiary functions, and changes may take place in organs by the occasional predominance of a subsidiary function over the original primary one. Thus the swim- or air-bladder which grows out dorsally from the food-canal of most fishes, seems usually to be a hydrostatic organ; in a few cases it helps slightly in respiration, but in the double-breathing mud-fishes or Dipnoi it has become a genuine lung. An unimportant (allantoic) bladder at the hind end of the gut in frogs, is represented in the embryos of reptiles and birds by a very important respiratory (and sometimes yolk-absorbing) birth-robe, and in almost all mammals by part of the placenta which unites mother and unborn offspring.

Substitution of Organs.—To the embryologist Kleinenberg we owe a suggestive conception of organic change, which he speaks of as the development of organs by substitution: An organ may supply the stimulus and the necessary condition for another which gradually supersedes and replaces it. In the simplest backboned animals, such as the lancelet, there is a supporting gristly rod along the back; among fishes the same rod or notochord is largely replaced by a backbone; in yet higher Vertebrates the adults have almost no notochord, its replacement by the backbone is almost complete. So in the individual life-history, all vertebrate embryos have a notochord to begin with; in the lancelet and some others this is retained throughout life, in higher forms it is temporary and serves as a scaffolding around which, from a thoroughly distinct embryological origin, the backbone develops. What is the relation between these two structures — notochord and backbone? According to Kleinenberg, the notochord supplies the necessary stimulus or condition for the development of the backbone which replaces it.

Rudimentary Organs. — (*a*) Through some ingrained defect it sometimes happens that an organ does not develop perfectly. The heart, the brain, the eye may be spoilt in the making. Such cases are illustrations of arrested development. (*b*) A parasitic crustacean, such as the *Sacculina* which shelters beneath the tail of a crab, begins life with many equipments such as legs, food-canal, eye, and brain, which are afterwards entirely or nearly lost; the sedentary adult sea-squirt or ascidian has lost the tail, the notochord, the spinal cord which its free-swimming tadpole-like larva possessed. Such cases are illustrations of degeneration. In these instances the retrogression is demonstrable in each lifetime, in other cases we have to compare the animal with its ancestral ideal. Thus there are many cave-animals whose eyes are always blind and abortive. The little kiwi of New Zealand has only apologies for wings. We need have no hesitation in calling these

animals degenerate in eyes and fore-limbs respectively.
(c) But somewhat different are such structures as the
following : The embryonic gill-clefts of reptiles, birds, and
mammals, which have no respiratory significance, or the
embryonic teeth of whalebone whales, of some parrots and
turtles, which in no case come to anything. They are
vestigial structures, which are partly explained on the
assumption, justified also in other ways, that the ancestors
of reptiles, birds, and mammals used the gill-clefts as fishes
and tadpoles do, that the ancestors of whalebone whales,
birds, and turtles had functional teeth. No one can say
with certainty of vestigial structures that they are entirely
useless, nor can one precisely say why they persist after
their original usefulness has ceased. They remain because
of necessities of growth of which we are ignorant, and
they may be useful in relation to other structures though
in themselves functionless.

Classification of Organs.—We may arrange organs
according to their work, some, such as limbs and weapons,
being busied with the external relations of the organism ;
others, such as heart and liver, being concerned with
internal affairs. Or we may classify them according to
their development from the outer, middle, or inner layer of
the embryo. Thus brain and sense-organs are always mainly
due to the outer stratum (ectoderm or epiblast), muscles
and skeleton arise from the middle mesoderm or mesoblast,
the gut and its outgrowths such as lungs and liver primarily
originate from the inner endoderm or hypoblast. Or we
may arrange the various structures more or less arbitrarily
for convenience of description as follows : the skin and its
outgrowths, appendages, skeleton, muscular system, nervous
system, sense-organs, the food-canal and its outgrowths,
the body-cavity, the heart and blood-vessels, the respiratory
organs, the excretory system, the reproductive organs.

Tissues.—To the school of Cuvier we owe the analysis
of the animal organism into its component organs ; but as
early as 1801 Bichât published his *Anatomie Générale*, in
which the analysis was carried a step farther. He reduced
the organs to their component tissues, and maintained that

the function of an organ might be expressed in terms of the properties of its tissues.

If we pass to the next step of analysis, and think of the body as a complex city of cells, we are better able to understand what tissues are. Each cell corresponds to a house, a tissue corresponds to a street of similar houses. In a city like Leipzig many streets are homogeneous, formed by houses or shops in which the predominant activity is the same throughout. A street is devoted to the making of clothes, or of bread, or of books. So in the animal body aggregates of contractile cells form muscular tissue, of supporting cells skeletal tissue, of secreting cells glandular tissue, and so on.

It is enough to state the general idea that a tissue is an aggregate of more or less similar cells, and to note that the different kinds may be grouped as follows:—

I. Nervous tissue, consisting of cells which receive, transmit, or originate nerve-stimuli.
II. Muscular tissue, consisting of contractile cells.
III. Epithelial tissue, consisting of lining and covering cells, which often become glandular, exuding the products of their activity as secretions.
IV. Connective tissue, including cells which bind, support, and store.

Cells.—To the discovery and perfecting of the microscope we owe the analysis of the body into its unit masses of living matter or cells. From 1838-39, when Schwann and Schleiden stated in their "cell doctrine" that all organisms—plants and animals alike—were built up of cells, cellular biology may be said to date. It was soon shown as a corollary that every organism which reproduced in the ordinary fashion arose from a single egg-cell or ovum which had been fertilised by union with a male-cell or spermatozoon. Moreover, the position of the simplest animals and plants was more clearly appreciated; they are single cells, the higher organisms are multicellular.

Now the cells of the animal body are necessarily varied, for the existence of a body involves division of labour among the units. Some, such as the lashed cells lining the windpipe, are very active, like the Infusorian Protozoa; others, for instance the fat-cells and gristle-cells of connective tissue, are very passive, something like the Gregarines; others, such as the white blood corpuscles or leucocytes, are between these extremes, and resemble the amœboid Protozoa.

But it is true of most of them that they consist (1) of a

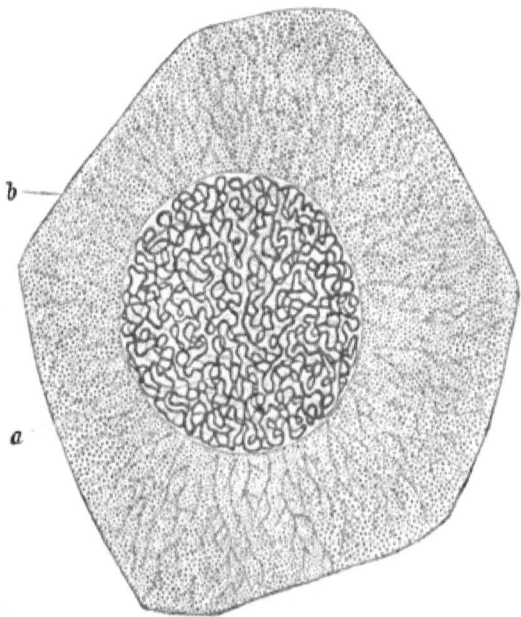

FIG. 34.—Animal cell, showing the coiled chromatin threads of the nucleus (*a*), and the protoplasmic network (*b*) round about. (From *Evolution of Sex*; after Carnoy.)

complex, and in part living cell-substance, in which keen eyes looking through good microscopes detect an intricate network, or sometimes the appearance of a fine foam; (2) of a central kernel or nucleus, which plays an important but hardly definable part in the life of the cell, especially during the process of cell-division; (3) of a slight outer membrane, varying much in definiteness and sometimes quite absent, through which communications with neighbouring cells are often established; and (4) of cell contents, which can be chemically analysed, and which are products of the vital activity rather than parts of the living substance, such as pigment, fat, and glycogen or animal starch.

The growth of all multicellular animals depends upon a multiplication of the component cells. Like organisms, cells have definite limits of growth which they rarely exceed; giants among the units are rare. When the limit of growth is reached the cell divides.

The necessity for this division has been partly explained by Spencer and Leuckart. If you take a round lump of dough, weighing an ounce, another of two ounces, a third of four ounces, you obviously have three masses successively doubled, but in doubling the mass you have not doubled the surface. The mass increases as the cube, the surface only as the square of the radius. Suppose these lumps alive, the second has twice as much living matter as the first, but not twice the surface. Yet it is through the surface that the living matter is fed, aerated, and purified. The unit will therefore get into physiological difficulties as it grows bigger, because its increase of surface does not keep pace with its increase of mass. Its waste tends to exceed its repair, its expenditure gains on its income. What are the alternatives? It may go on growing and die (but this is not likely), it may cease growing at the fit limit, it may greatly increase its surface by outflowing processes (which thus may be regarded as life-saving), or it may divide. The last is the usual course. When the unit has grown as large as it can conveniently grow, it divides; in other words, it reproduces at the limit of growth, when processes of waste are gaining on those of construction. By dividing, the mass is lessened, the surface increased, the life continued.

But although we thus get a general rationale of cell-division, we are not much nearer a conception of the internal forces which operate when a cell divides; for in most cases the process is orderly and complex, and is somehow governed by the behaviour of the nucleus. Few results of the modern study of minute structure are more marvellous than those which relate to dividing cells. From Protozoa to man, and also in plants, the process is strikingly uniform. The nucleus of the cell becomes more active, the coil or network of threads which it contains is undone and takes the new and more regular form of a spindle or barrel. The division is most thorough, each of the two daughter-cells getting an accurate half of the original nucleus. Recent investigators, moreover, assert that from certain centres in the cell-substance an influence is exerted on the nuclear threads, and they talk of an archoplasm within the protoplasm, and of marked individuality of behaviour in the nuclear threads.

From the cell as a unit element we penetrate to the protoplasm which makes it what it is. Within this we discern an intricate network, within this, special centres of force—"attractive spheres" and "central corpuscles," or an "archoplasm" within the protoplasm! We study the nucleus, first as a simple unit which divides, years afterwards as composed of a network or coil of nuclear threads which seem ever to become more and more marvellous, "behaving like little organisms." We split these up into "microsomata," and so on, and so on. But we do not catch the life of the cell, we cannot locate it, we cannot give an account of the mechanics of cell-division. It is a mystery of life. After all our analysis we have to confess that the cell, or the protoplasm, or the archoplasm, or the chromatin threads of the nucleus, or the "microsomata" which compose them, baffle our analysis; they behave as they do because they are alive. Were we omniscient chemists, such as Laplace imagined in one of his speculations, and knew the secret of protoplasm, how its touch upon the simpler states of matter is powerful to give them life, we should but have completed a small part of those labours that even now lie waiting us; what further investigations will present themselves we cannot tell.

CHAPTER XII

THE LIFE-HISTORY OF ANIMALS

1. *Modes of Reproduction*—2. *Divergent Modes of Reproduction*—3. *Historical*—4. *The Egg-Cell or Ovum*—5. *The Male-Cell or Spermatozoon*—6. *Maturation of the Ovum*—7. *Fertilisation*—8. *Segmentation and the first stages in Development*—9. *Some Generalisations—The Ovum Theory, the Gastræa Theory, Fact of Recapitulation, Organic Continuity*

IN his exercitation "on the efficient cause of the chicken," Harvey (1651) confesses that "although it be a known thing subscribed by all, that the fœtus assumes its original and birth from the male and female, and consequently that the egge is produced by the cock and henne, and the chicken out of the egge, yet neither the schools of physicians nor Aristotle's discerning brain have disclosed the manner how the cock and his seed doth mint and coine the chicken out of the egge." The marvellous facts of growth are familiar to us — the sprouting corn and the opening flowers, the growth of the chick within the egg and of the child within the womb; yet so difficult is the task of inquiring wisely into this marvellous renewal of life that we must reiterate the old confession: "ingratissimum opus scribere ab iis quæ, multis a natura circumjectis tenebris velata, sensuum lucis inaccessa, hominum agitantur opinionibus."

1. **Modes of Reproduction.**—The simplest animals divide into two or into many parts, each of which becomes a full-grown Protozoon. There is no difficulty in understanding

why each part should be able to regrow the whole, for each is a fair sample of the original whole. Indeed, when a large Protozoon is cut into two or three pieces with a knife, each fragment is often able to retain the movements and life of the intact organism. Among the Protozoa we find some in which the multiplication looks like the rupture of a cell which has become too large; in others numerous buds are set free from the surface; in others one definitely-formed bud (like an overflow of the living matter) is set free; in others the cell divides into two equal parts, after the manner of most cells; and numerous divisions may also occur in rapid succession and within a cyst, that is, in limited time and space, with the result that many "spores" are formed. These modes of multiplication form a natural series.

In the many-celled animals multiplication may still proceed by the separation of parts; indeed the essence of reproduction always is the separation of part of an organism to form—or to help to form—a new life. Sponges bud profusely, and pieces are sometimes set adrift; the *Hydra* forms daughter polypes by budding, and these are set free; sea-anemones and several worms, and perhaps even some star-fishes, multiply by the separation of comparatively large pieces. But this mode of multiplication—which is called asexual—has evident limitations. It is an expensive way of multiplying. It is possible only among comparatively simple animals in which there is no very high degree of differentiation and integration. For though cut-off pieces of a sponge, *Hydra*, sea-anemone, or simple worm may grow into adult animals, this is obviously not the case with a lobster, a snail, or a fish. Thus with the exception of the degenerate Tunicates there is no budding among Vertebrates, nor among Molluscs, nor among Arthropods.

The asexual process of liberating more or less large parts, being expensive, and possible only in simpler animals, is always either replaced or accompanied by another method—that of sexual reproduction. The phrase "sexual reproduction" covers several distinct facts: (*a*) the separation of special reproductive cells; (*b*) the production of two different kinds of reproductive cells (spermatozoa and ova), which are dependent on one another, for in most cases an ovum comes to nothing unless it be united with a male-cell or spermatozoon, and in all cases the spermatozoon comes to nothing unless it be united with an ovum; (*c*) the production of spermatozoa and ova by different (male and female) organs or individuals.

(*a*) It is easy to think of simple many-celled animals being multiplied by liberated reproductive cells, which differed but little from those of the body. But as more and more division of labour was established in the bodies of animals, the distinctness of the reproductive cells from the other units of the body became greater. Finally, the prevalent state was reached, in which the only cells able to begin a new life when liberated are the reproductive cells. They owe this power to the fact that they have not shared in making the body, but have preserved intact the characters of the fertilised ovum from which the parent itself arose.

(*b*) But, in the second place, it is easy to conceive of a simple multicellular animal whose liberated reproductive cells were each and all alike able to grow into new organisms. In such a case, we might speak of sexual reproduction in one sense, for the process would be different from the asexual method of liberating more or less large parts. But yet there would be no fertilisation and no sex, for fertilisation means the union of mutually dependent reproductive cells, and sex means the existence of two physiologically different kinds of individuals, or at least of organs producing different kinds of reproductive cells. We can infer from the Protozoa how fertilisation or the union of the two kinds of reproductive cells may have had a gradual origin. For in some of the simplest Protozoa, *e.g. Protomyxa*, a large number of similar cells sometimes flow together; in a few cases three or more combine; in most a couple of apparently similar units unite; while in a few instances, *e.g. Vorticella*, a small cell fuses with a large one, just as a spermatozoon unites with an ovum.

(*c*) But the higher forms of sexual reproduction imply more than the liberation of special reproductive cells, more than the union of two different and mutually dependent kinds of reproductive cells, — they imply the separation of the sexes. The problem of sexual reproduction becomes less difficult when the various facts are discussed separately, and if you grant that there is no great difficulty in understanding the liberation of special cells, and no great difficulty in understanding why two different kinds should in most cases have to unite if either is to develop, then I do not think that the remaining fact — the evolution of male and female individuals—need remain obscure.

If we study those interesting Infusorian colonies, of which *Volvox* is a good type, the riddle may be at least partially read. Though Protozoa, they are balls of cells, in which the component units are united by protoplasmic bridges and show almost no division of labour. From such a ball of cells, units are sometimes set free which divide and form new colonies. In other conditions a less direct multiplication occurs. Some of the cells—apparently better fed than their neighbours—become large; others, less successful, divide into many minute units. The large kind of cell is fertilised by the small kind of cell, and there is no reason why we should not call them ova and spermatozoa respectively. In such a *Volvox*, two different kinds of reproductive cell are made within one organism. But we also find *Volvox* balls in which only ova are being made, and others in which only spermatozoa are being made. The sexes are separate. Indeed we have in *Volvox*, as Dr. Klein—an enthusiastic investigator of this form —rightly says, an epitome of all the great steps in the evolution of sex.

So far I have stated facts; now I shall briefly state the theory by which Professor Geddes has sought to rationalise these facts.

All through the animal series, from the active Infusorians and passive Gregarines, to the feverish birds and sluggish reptiles, and down into the detailed contrasts between order and order, species and species, an antithesis may be read between predominant activity and preponderant passivity, between lavish expenditure of energy and a habit of storing, between a relatively more disruptive (*katabolic*) and a relatively more constructive (*anabolic*) series of changes in the protoplasmic life of the creature. The contrast between the sexes is an expression of this fundamental alternative of variation.

The theory is confirmed by contrasting the characteristic product of female life—passive ova, with the characteristic product of male life—active spermatozoa ; or by summing up the complex conditions (abundant food, favourable temperature, and the like) which favour the production of female offspring, with the opposite conditions which favour maleness ; or by contrasting the secondary sexual characters of the more active males (*e.g.* bright colours, smaller size) with the opposite characteristics of their more passive mates.

Apart from the general problem of the evolution of sex, those who find the subject interesting should think about the evolution of the so-called "sexual instincts," as illustrated in the attraction of mate to mate. As to the actual facts of pairing and giving birth, it seems to me that I have suggested the most profitable way of considering these in a former part of this book where courtship and parental care are discussed, though I believe firmly with Thoreau, that "for him to whom sex is impure, there are no flowers in nature."

2. **Divergent Modes of Reproduction.**—(*a*) *Hermaphroditism.*—Especially among lower animals, both ova and spermatozoa may be produced by one individual, which is then said to be hermaphrodite. So most common plants produce both seeds and pollen. Some sponges and stinging animals, many "worms," *e.g.* earthworm and leech, barnacles and acorn-shells among crustaceans, one of the edible oysters, the snail, and many other molluscs, the sea-squirts, and the hagfish, are all hermaphrodite. But it should be noted that the organs in which ova and spermatozoa are produced are in most cases separate, that the two

kinds of cells are usually formed at different times, and that the fertilisation of ova by spermatozoa from the same animal very rarely occurs. It is very likely that the bisexual or hermaphrodite state of periodic maleness and femaleness is more primitive than that of separate sexes, which, except in tunicates, a few fishes and amphibians, and casual abnormalities, is constant among the backboned animals.

(*b*) *Parthenogenesis* seems to be a degenerate form of sexual reproduction in which the ova produced by female organisms develop without being fertilised by male cells. Thus "the drones have a mother but no father," for they develop from ova which are not fertilised. In some rotifers the males have never been found, and yet the fertility of the females is very great; in many small crustaceans ("water-fleas") the males seem to die off and are unrepresented for long periods; in the aphides males may be absent for a summer (or in a greenhouse for years) without affecting the rapid succession of female generations.

(*c*) *Alternation of Generations.*—A fixed asexual zoophyte or hydroid sometimes buds off and liberates sexual swimming bells or medusoids, whose fertilised ova develop into embryos which settle down and grow into hydroids. This is perhaps the simplest and clearest illustration of alternation of generations.

In autumn the freshwater sponge (*Spongilla*) begins to suffer from the cold and the scarcity of food. It dies away; but some of the units club together to form "gemmules" from which in spring male and female sponges are developed. The males are short-lived, but their spermatozoa fertilise the ova of the females. The fertilised ovum develops into a ciliated embryo, and this into the asexual sponge, which produces the gemmules.

The large free-swimming and sexual jellyfishes of the genus *Aurelia* produce ova and spermatozoa; from the fertilised ovum an embryo develops not into a jellyfish, but into a sessile *Hydra*-like animal. This grows and divides and gives origin asexually to jellyfish.

Similar but sometimes more complicated alternations

occur in some worm-types (some flukes, threadworms, etc.), and as high up in the series as Tunicates; while among plants analogous alternations are very common, *e.g.* in the life-cycles of fern and moss.

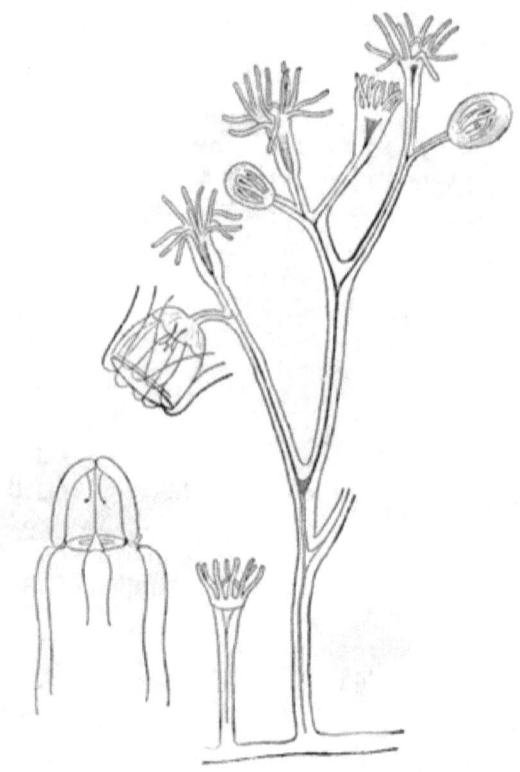

Fig. 35.—Diagram of a hydroid colony, some of the individuals of which have been modified as swimming-bells or medusoids; one of these has been liberated.

3. **Historical.**—In the seventeenth and eighteenth centuries, naturalists had a short and easy method of dealing with embryology. They maintained that within the seed of a plant, within the egg of a bird, the future organism was already present in miniature. Every germ contained a miniature model of the adult, which in development was

simply unfolded. It was to this unfolding that the word evolution (as a biological term) was first applied. But not only did they compare the germ to a complex bud hiding the already formed organs within its hull, they maintained that it included also the next generation and the next and the next. Some said that the ovum was most important, that it required only the sperm's awakening touch and it began to unfold; others said that the animalcules or spermatozoa produced by male animals were most important, that they only required to be nourished by the ova. The two schools nicknamed one another "ovists" and "animalculists." The preformation-theories were false, as Harvey in the middle of the seventeenth century discerned, and as Wolff a century later proved, because germs are demonstrably simple, and because embryos grow gradually part by part. But in a later chapter we shall see that the theories were also strangely true.

4. **The Egg-cell or Ovum** produced by a female animal, or at least by a female organ (ovary), exhibits the usual characteristics of a cell. It often begins like an Amœba, and may absorb adjacent cells; in most cases it becomes surrounded by an envelope or by several sheaths; in many cases it is richly laden with yolk derived from various sources. In the egg of a fowl, the most important part (out of which the embryo is made) is a small area of transparent living matter which lies on the top of the yellow yolk and has a nucleus for its centre; round about there is a coating of white-of-egg; this is surrounded by a double membrane which forms an air-chamber at the broad end of the egg; outermost is the porous shell of lime.

While there must be a general relation between the size of the bird and that of the egg, there are many inconsistencies, as you will soon discover if you compare the eggs of several birds of the same size. It is said that the eggs of birds which are rapidly hatched and soon leave the nest tend to be large, and that there is some relation between the size of eggs and the number which the bird has to cover. It seems probable, however, from what one notices in the poultry yard and in comparing the constitution of different birds, that a highly-nourished and not very energetic bird will have larger eggs than one of more active habits and sparser diet.

The egg-shell consists almost wholly of carbonate of lime, and the experiments of Irvine have shown that a hen can form a carbonate of lime shell from other lime salts. It is formed around the egg in the lower part of the oviduct, and is often beautifully coloured with pigments allied to those of blood and bile. These colours often harmonise well with the surroundings, but how this advantageous result has been wrought out is uncertain.

Eggs differ greatly in regard to the amount of yolk which they contain; thus those of birds and reptiles have much, while those of all mammals except the old-fashioned Monotremes have hardly any. This is related partly to the number of eggs which are produced, and partly to the amount of food-capital which the embryo requires before other sources of supply become available. The young of birds and reptiles feed on the yolk until they are hatched, the unborn young of all the higher (placental) mammals absorb food from the mothers. The different sizes of egg usually depend upon the amount of yolk, for the really vital portion out of which the embryo is made is always very small.

There are many differences also in regard to the outer envelopes, witness the jelly around the spawn of frogs, the firm but delicate skin around the ova of cuttlefish, the "horny" mermaid's-purse enclosing the skate's egg, the chitinous sheath surrounding the ova of many insects, the calcareous shell in birds and most reptiles.

5. **The Male-Cell or Spermatozoon** produced from a male animal, or at least from a male organ (testis), is very different from the ovum. It is very minute and very active. If we compare an ovum to an Amœba or to an encysted Gregarine among Protozoa, we may liken the spermatozoon to a minute monad Infusorian. It is a very small cell, bearing at one end a "head," which consists mostly of nucleus, prolonged at the other end into a mobile "tail," which lashes the head along.

The spermatozoon, though physiologically the complement of the ovum, is not its morphological equivalent. The precise equivalent of the ovum is a primitive male-cell or mother-sperm-cell, which divides repeatedly and forms a ball or clump of spermatozoa. This division is to be compared with the division or segmentation of the ovum, which we shall afterwards discuss.

In some cases spermatozoa which have been transferred to a female may lie long dormant there. Thus those received by the queen-bee during her nuptial flight may last for a whole season, or even for three seasons, during which they are used in fertilising those ova which develop into workers or queen-bees. Quite unique is the case of one of Sir John Lubbock's queen-ants, which, thirteen years after the last sexual union with a male, laid eggs which developed.

6. **Maturation of the Ovum.**—Most ova before they are fertilised are subject to a remarkable change, the precise meaning of which is not certainly known. The nucleus of the ovum moves to the surface and is halved twice in rapid succession. Two minute cells or polar globules are thus extruded, and come to nothing, while the bulk of the nucleus is obviously reduced by three-fourths. It may be that the ovum is only behaving as other cells do at the limit of growth, or that it is exhibiting in an ineffective sort of way the power of independent division which all the reproductive cells of very simple many-celled animals perhaps possessed; it may be that it is parting with some surplus material which is inconsistent with or no longer necessary to its welfare, and there are other theories. One fact, however, seems well established, that parthenogenetic ova, which are able to develop into embryos without being fertilised, extrude only one polar globule, a fact which suggests that the amount of nucleus thus retained somehow makes up for the absence of a spermatozoon.

7. **Fertilisation.**—When a pollen grain is carried by an insect or by the wind to the stigma of a flower, it grows down through the tissue of the pistil until it reaches the ovule and the egg-cell which that contains. Then a nuclear

element belonging to the pollen cell unites with the nucleus of the egg-cell. The union is intimate and complete.

When spermatozoa come in contact with the egg-shell of a cockroach ovum, they move round and round it in varying orbits until one finds entrance through a minute aperture in the shell. It works its way inwards until its nuclear part unites with that of the ovum. The union is again intimate and complete.

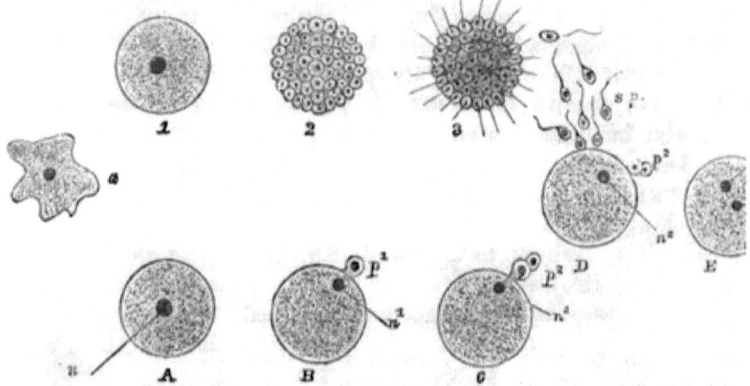

FIG. 36.—Diagram of the development of spermatozoa (upper line), of the maturation and fertilisation of the ovum (lower line).

a, primitive amœboid sex-cell; A, ovum with nucleus (n); B, ovum extruding the first polar body (p^1) and leaving the nucleus (n^1) reduced by half; C, extrusion of the second polar body (p^2), the nucleus (n^2) now reduced to a fourth of its original size; 1, a mother-sperm-cell, dividing (2 and 3) into spermatozoa (sp); D, the entrance of a spermatozoon into the ovum; E, the male nucleus ($sp.n$) and the female nucleus (n^2) approach one another, and are about to be united, thus consummating the fertilisation. (From the *Evolution of Sex*.)

Both in plants and in animals the male cell is attracted to the female cell, the two nuclei unite thoroughly, and, when fertilisation is thus effected, the egg-cell is usually impervious to other sperms.

A single nucleus of double origin is thus established, and the egg-cell begins to divide. Some idea both of the orderly complexity of the nuclear union and of the carefulness of modern investigation may be gained from the fact that the nuclei of the two daughter-cells which result from the first division of the egg-cell have been shown to consist in equal proportions of material derived from the male-nucleus and from the ovum-nucleus.

Yet in the last century naturalists still spoke of an "aura seminalis," and believed that a mere breath, as it were, of the male cell was sufficient to fertilise an egg, and it was only in 1843 that Martin Barry discerned the presence of the spermatozoon within the ovum.

8. **Segmentation and Development.** — The fertilised egg-cell divides, and by repeated division and growth of cells every embryo, of herb and tree, of bird and beast, is formed. On the quantity and arrangement of the yolk the character of the segmentation depends. When there is little or no yolk the whole ovum divides into equal parts, as in sponge, earthworm, starfish, lancelet, and higher mammal. When there is more than a little yolk, and when this sinks to the lower part of the egg-cell, the division is complete but unequal, and this may be readily seen by examining freshly-laid frog spawn. When the yolk is accumulated in the core of the egg-cell, the more vital superficial part divides, as in insects and crustaceans. Lastly, when the yolk is present in large quantity as in the ova of gristly fishes, reptiles, and birds, the division is very partial, being confined to a small but rapidly extending area of formative living matter, which lies like a drop on the surface of the yolk.

As the result of continued division, a ball of cells is formed. This may be hollow (a *blastosphere*), or solid (a *morula, i.e.* like a mulberry), or it may be much modified in form by the presence of a large quantity of yolk. Thus in the hen's egg what is first formed is a disc of cells technically called the *blastoderm*, which gradually spreads around the yolk.

The hollow ball of cells almost always becomes dimpled in or invaginated, as an india-rubber ball with a hole in it might be pressed into a cup-like form. The dimpling is the result of inequalities of growth. The two-layered sac of cells which results is called a *gastrula*, and the cavity of this sac becomes in the adult organism the digestive part of the food-canal. Where there is no hollow ball of cells, but some other result of segmentation, the formation of a gastrula is not so obvious. Yet in most cases some analogous infolding is demonstrable.

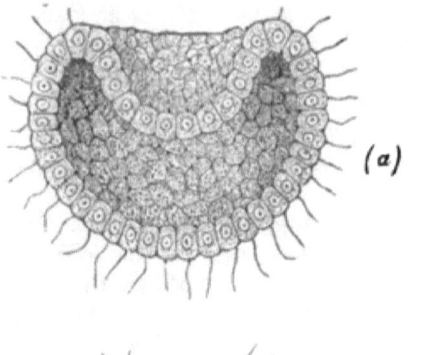

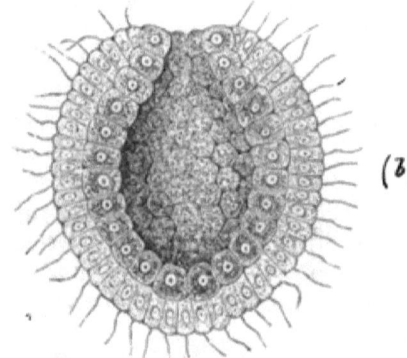

FIG. 37.—The formation of the two-layered gastrula from the invagination of a hollow sphere of cells. (From the *Evolution of Sex*; after Haeckel.)

In the hollow sac of cells there are already two layers. The outer, which is called the ectoderm or epiblast, forms in the adult the outer skin, the nervous system, and the most important parts of the sense-organs. The inner, which is called the endoderm or hypoblast, forms the lining of the most important part of the food-canal, and of such appendages as lungs, liver, and pancreas which are outgrowths from it. But in all animals above the Sponges and Cœlenterates, a middle layer appears between the other two. From this—the mesoderm or mesoblast—the muscles, the internal skeleton, the connective-tissue, etc., are formed.

9. **Some Generalisations.**—(*a*) *The "Ovum-Theory."* To realise that almost every organism from the sponge to the highest begins its life as a fertilised egg-cell, and is built up by the division and arrangement, layering and folding of cells, should not lessen, but should greatly enhance, the wonder with which we look upon life. If the end of this constantly repeated process of development be

something to marvel at, the same is equally true of its beginning.

(*b*) *The Gastræa Theory.* From the frequent, though not universal occurrence of the two-layered gastrula stage in the development of animals, Haeckel concluded that the first stable form of many-celled animal must have been something very like a gastrula. He called this hypothetical ancestor of all many-celled animals a *Gastræa*, and his inference has found favour with many naturalists. Some of the simplest sponges, polypes, and "worms" are hardly above the gastrula level.

(*c*) *Recapitulation.* When we take a general survey of the animal series, we recognise that the simplest animals are single cells, that the next simplest are balls of cells like *Volvox*, and that the next simplest are two-layered sacs of cells like the simple sponges, polypes, and worms above referred to. These represent the three lowest steps in the evolution of the race. They are not hypothetical steps in a hypothetical ladder of ascent, they are realities.

When we take a general survey of the individual development of many-celled animals, we recognise that all begin as single egg-cells, and that the ova divide into balls of cells, which become in most cases two-layered sacs of cells. It is therefore evident that the first three chapters in individual history are precisely the first three steps in racial history.

Von Baer, one of the pioneer embryologists in the first half of this century, discerned that the individual life-history was in its general course a recapitulation of the history of the race. He recognised that even one of the higher animals, let us say a rabbit, began at the beginning as a Protozoon, that it slowly acquired the features of a primitive Vertebrate, that it subsequently showed the character of a young fish, afterwards of a young reptile, then of a young mammal, then of a young rodent, finally of a young rabbit. He confessed his inability to distinguish whether three very young embryos, freed from their surroundings, were those of reptiles, birds, or mammals. In stating Von Baer's vivid idea of development as progress from the simple to the complex, from the general to the special, we must be careful to notice that he did not say that the young mammal was once like a little fish, afterwards like a reptile, and so on; he compared the embryo mammal at one stage with the embryo fish, at another stage with the embryo reptile, which is a very different matter.

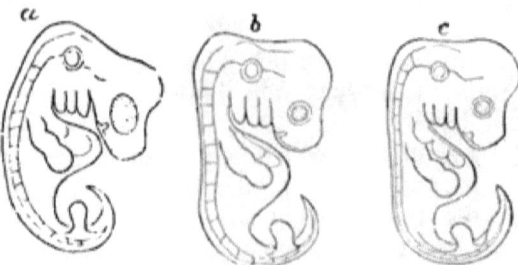

FIG. 38.—Embryos of fowl, *a*; dog, *b*; man, *c*. (From Chambers's *Encyclop.*; after Haeckel.)

Fritz Müller, in his *Facts for Darwin*, illustrated the same idea in relation to Crustacea. When a young crayfish is hatched, it is practically a miniature adult. When a young lobster is hatched, it differs not a little from the adult, and is described as being at a *Mysis* stage,—*Mysis* being a prawn-like crustacean. It grows and moults and becomes a little lobster. When a crab is hatched, it is quite unlike the adult, it is liker one of the humblest Crustacea such as the common water-flea *Cyclops*, and is described as a Zoea. This Zoea grows and moults and becomes, not yet a crab but a prawn-like animal with extended tail, a stage known as the *Megalopa*. This grows and moults, tucks in its tail, and becomes a young crab. And again, when the shrimp-like crustacean, known as *Penæus*, is hatched, it is simpler than any known crustacean, it is an unringed somewhat shield-shaped little creature with three pairs of appendages and a median eye. It is known as a Nauplius and resembles the larvæ of most of the simpler crustaceans. It grows and moults and becomes a Zoea, grows and moults and becomes a *Mysis*, grows and moults and becomes a *Penæus*.

Now these life-histories are hardly intelligible at all unless we believe that *Penæus* does in some measure recapitulate the steps of racial progress, that the crab does so to a slighter extent, that the lobster has abbreviated its obvious recapitulation much more, while the crayfish has found out a short cut in development. Let us exercise our imagination and think of the ancestral Crustacea perhaps not much less simple than the Nauplius larvæ which many

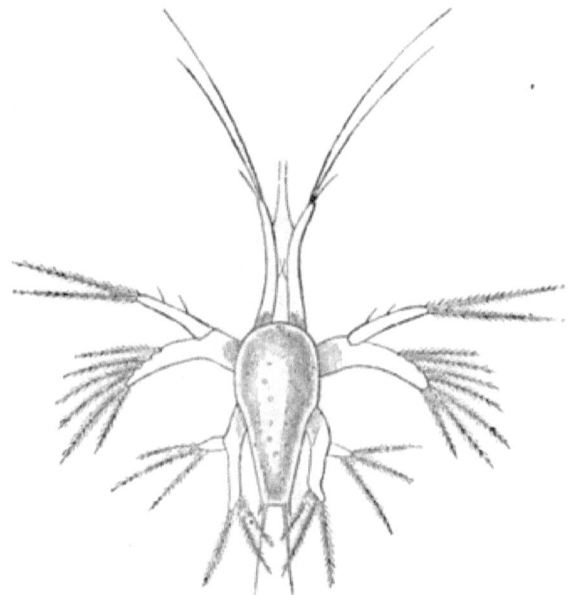

FIG. 39.—Life-history of *Penæus*; the Nauplius.

of them exhibit. In the course of time some pushed forward in evolution and attained to the level of structure represented by the Zoea larvæ. At this station some remained and we have already mentioned the "water-flea" *Cyclops* as a crustacean which persists near this level. But others pushed on and reached a stage represented by *Mysis*, and finally the highest crustaceans were evolved.

Now to a certain extent these highest crustaceans have to travel in their individual development along the rails laid down in the progress of the race. Thus *Penæus*, starting of course as an ovum at the level of the Protozoa, has to stop as it were at the first distinctively crustacean station—the Nauplius stage. After some change and delay, it continues to progress, but again there is a halt and a change at the Zoea station. Finally there is another

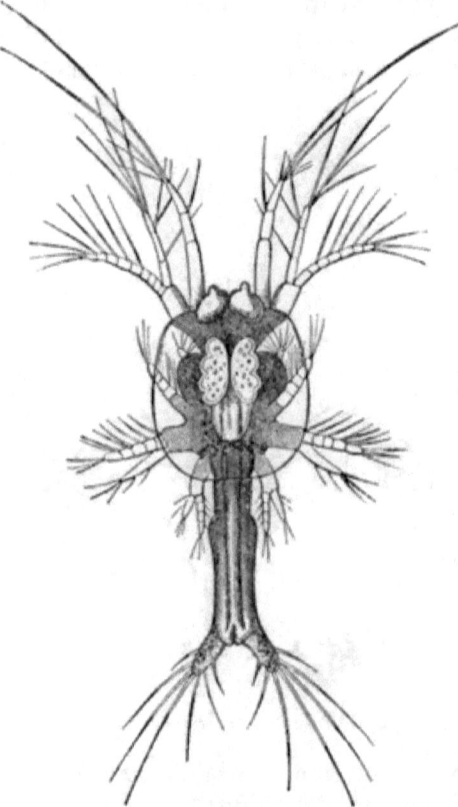

FIG. 39*a*.—Life-history of *Penæus*; the Zoea.

delay at the *Mysis* stage before the *Penæus* reaches its destination. The crab, on the other hand, stops first at the Zoea station, the lobster at the *Mysis* station, while the crayfish though progressing very gradually like all the others has—if you do not find the simile too grotesque—a through-carriage all the way.

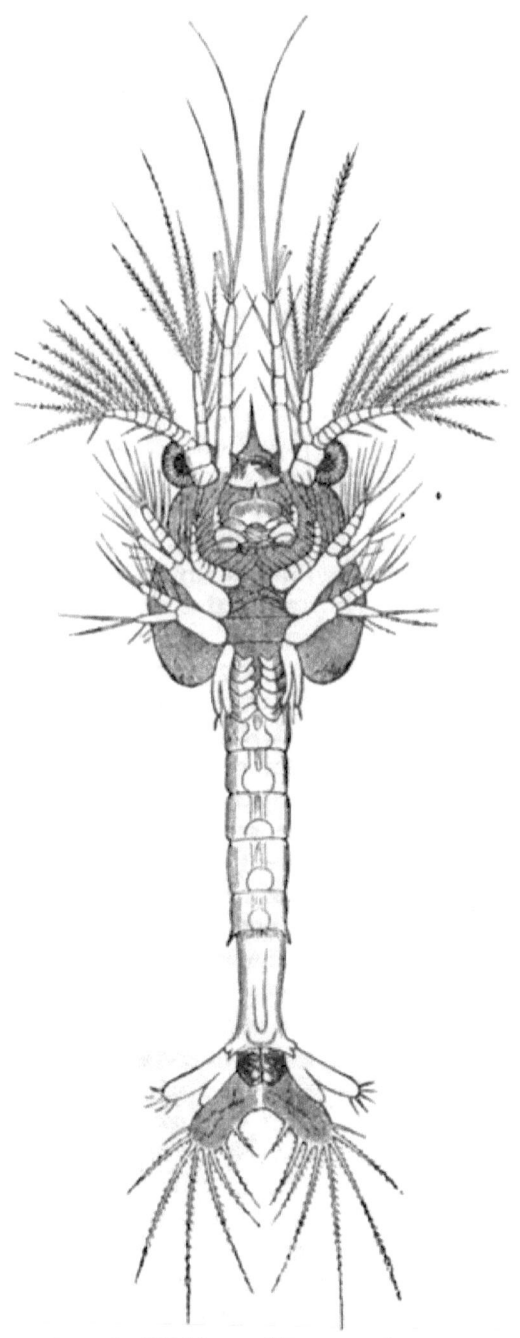

Fig. 39b.—Life-history of *Penæus*; a later stage.

One must be careful not to press the idea of recapitulation too far, (1) because the individual life-history tends to skip stages which occurred in the ancestral progress; (2) because the young animal may acquire new characters which are peculiar to its own near lineage and have little or no importance in connection with the general evolution of its race; (3) because, in short, the resemblance between the individual and racial history (so far as we know them) is general, not precise. Thus we regard Nauplius and Zoea rather as adaptive larval forms than as representatives of ancestral crustaceans. Moreover, if one insists too much on the approximate parallelism between the life-history of the individual and the progress of the race, one is apt to overlook the deeper problem — how it is that the recapitulation occurs to the extent that it undoubtedly does. The organism has no feeling for history that it should tread a sometimes circuitous path, because its far-off ancestors did so. To some extent we may think of inherited constitution as if it were the hand of the past upon the organism, compelling it to become thus or thus, but we must realise that this is a living not a dead hand; in other words these metamorphoses have their efficient causes in the actual conditions of growth and development. The suggestion of Kleinenberg referred to in a preceding chapter helps us, for

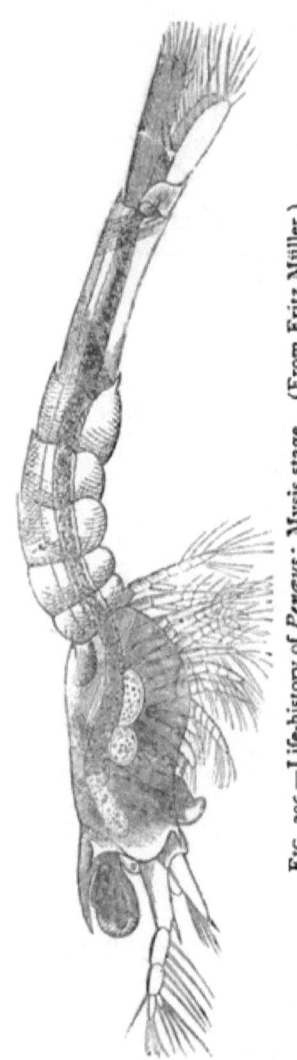

FIG. 39c.—Life-history of *Penæus*; Mysis stage. (From Fritz Müller.)

if we ask why an animal develops a notochord only to have it rapidly replaced by a backbone, part of the answer surely is that the notochord which in the historical evolution supplied the stimulus necessary for the development of a backbone, is still necessary in the individual history for the same purpose.

But there is no doubt that the idea of recapitulation is a very helpful one, in regard to our own history as well as in regard to animals, and we would do well to think of it much, and to read how Herbert Spencer (*Principles of Biology*, Lond. 1864-66) has discussed it in harmony with his general formula of evolution as a progress from the homogeneous to the heterogeneous; how Haeckel (*Generelle Morphologie*, Berlin, 1866) has illustrated it, and pithily summed it up in his "fundamental law of biogenesis" (*Biogenetisches Grundgesetz*), saying that ontogeny (individual development) recapitulates phylogeny (racial history); how Milnes Marshall (see *Nature*, Sept. 1890) has recently tested and criticised it, defining the limits within which the notion can be regarded as true, and searching for a deeper rationale of the facts than the theory supplies.

(*d*) *Organic Continuity*. In a subsequent chapter on heredity, which simply means the relation of organic continuity between successive generations, I shall explain the fundamental idea that the reproductive cells owe their power of developing, and of developing into organisms like the parents, to the fact that they are in a sense continuous with those which gave origin to the parents. A fertilised egg-cell with certain qualities divides and forms a "body" in which these qualities are expressed, distributed, and altered in many ways by division of labour. But it also forms reproductive cells, which do not share in the up-building of the body, which are reproductive cells in fact because they do not do so, because they retain the intrinsic qualities of the original fertilised ovum, because they preserve its protoplasmic tradition. If this be so, and there is much reason to believe it, then it is natural and necessary that these cells, liberated in due time, should behave as those behaved whose qualities they retain. It is necessary that like should beget like.

CHAPTER XIII

THE PAST HISTORY OF ANIMALS

1. *The two Records*—2. *Imperfection of the Geological Record*—3. *Palæontological Series*—4. *Extinction of Types*—5. *Various Difficulties*—6. *Relative Antiquity of Animals*

1. **The Two Records.**—Reviewing the development of the chick, W. K. Parker said, "Whilst at work I seemed to myself to have been endeavouring to decipher a palimpsest, and that not erased and written upon just once, but five or six times over. Having erased, as it were, the characters of the culminating type—those of the gaudy Indian bird—I seemed to be amongst the sombre grouse, and then, towards incubation, the characters of the Sand-Grouse and Hemipod stood out before me. Rubbing these away, in my downward walk, the form of the Tinamou looked me in the face; then the aberrant Ostrich seemed to be described in large archaic characters; a little while and these faded into what could just be read off as pertaining to the Sea Turtle; whilst, underlying the whole, the Fish in its simplest Myxinoid form could be traced in morphological hieroglyphics."

There is another palimpsest—the geological record written in the rocks. For beneath the forms which disappeared, as it were, yesterday,—the Dodo and the Solitaire, the Moa and the Mammoth, the Cave Lion and the Irish Elk,—there are mammals and birds of old-fashioned type the like of which no longer live. Beneath these lie the giant reptiles, beneath these great amphibians, preceded by hosts of armoured fishes, beyond the first traces of which only backboneless animals are found. Yet throughout the chapters of this record, written during different æons on the earth's surface, persistent forms recur from age to age, many of them, such as some of the lamp-shells or Brachiopods, living on from near the apparent beginning even until now. But other races, like the Trilobites, have died out, leaving none which we can regard as in any sense their direct descendants. Other sets of animals, like the Ganoid fishes, grow in strength, attain a golden age of prosperous success, and wane away. As the earth grew older nobler forms appeared, and this history from the tombs, like that from the cradles of animals, shows throughout a gradual progress from simple to complex.

2. Imperfection of the Geological Record.—If complete records of past ages were safely buried in great treasure-houses such as Frederic Harrison proposes to make for the enlightenment of posterity, then palæontology would be easy. Then a genealogical tree connecting the Protist and Man would be possible, for we should have under our eyes what is now but a dream—a complete record of the past.

The record of the rocks is often compared to a library in which shelves have been destroyed and confused, in which most of the sets of volumes are incomplete, and most of the individual books much damaged. When we consider the softness of many animals, the chances against their being entombed, and the history of the earth's crust, our wonder is that the record is so complete as it is, that from "the strange graveyards of the buried past" we can learn so much about the life that once was.

We must not suppose the record to be as imperfect as our knowledge of it. Thus many regions of the earth's surface have been very partially studied, many have not been explored at all, many are inaccessible beneath the sea.

As to the record, the rocks in which fossils are found are sedimentary rocks formed under water, often they have been unmade and remade, burnt and denuded. The chances against preservation are many.

Soft animals rarely admit of preservation, those living on land and in the air are much less likely to be preserved than those living in water, the corpses of animals are often devoured or dissolved. Again the chances against preservation are many.

3. **Palæontological Series.**—Imperfect as the geological record is, several marvellously **complete** series of related animals have been disentombed. Thus, a series of fossilised freshwater snails (*Planorbis*) has been carefully worked out; its extremes are very different, but the distinctions between any two of the intermediate forms are hardly perceptible. The same is true in regard to another set of freshwater snails (*Paludina*), and on a much larger scale among the extinct cuttlefishes (Ammonites, etc.) whose shells have been thoroughly preserved. The modern crocodiles are linked by many intermediate forms to their extinct ancestors, and the modern horse to its pigmy progenitors. In cases like these, the evidences of continuously progressive evolution are conclusive.

4. **Extinction of Types.**—A few animals, such as some of the lamp-shells or Brachiopods, have persisted from almost the oldest rock-recorded ages till now. In most cases, however, the character of the family or order or class has gradually changed, and though the ancient forms are no longer represented, their descendants are with us. There is an extinction of individuals and a slow change of species.

On the other hand there are not a few fossil animals which have become wholly extinct, whose type is not represented in the modern fauna. Thus there are no animals alive that can be regarded as the lineal descendants of Trilobites and Eurypterids, or of many of the ancient reptiles. There is no doubt that a race may die out. Many different kinds of heavily armoured Ganoid fishes abounded in the ages when the Old Red Sandstone was formed, but only seven different kinds are now alive. The lamp-shells and the sea-lilies, once very numerous, are now greatly restricted. Once there were giants among Amphibians, now almost all are pigmies.

It is difficult to explain why some of the old types disappeared. The extinction was never sudden. Formidable competitors may have helped to weed out some; for cuttlefish would tend to exterminate trilobites, and voracious fishes would decimate cuttlefish, just as man himself is rapidly and inexcusably annihilating many kinds of beasts and birds. But, apart from the struggle with competitors, it is likely that some types were insufficiently plastic to save themselves from changes of environment, and it seems likely that others were victims to their own constitutions, becoming too large, or too sluggish, or too calcareous; or, on the other hand, too feverishly active. The "scouts" of evolution would be apt to become martyrs to progress; the "laggards" in the race would tend to become pillars of salt; the path of success was oftenest a *via media* of compromise. Samuel Butler has some evidence for saying that "the race is not in the long run to the phenomenally swift, nor the battle to the phenomenally strong; but to the good average all-round organism that is alike shy of radical crotchets and old world obstructiveness."

5. **Various Difficulties.**—Nowadays it seems natural to us to regard the fossils in the rocks as vestiges of a gradual progress or evolution. As some still find difficulty in accepting this interpretation, I shall refer to three difficulties occasionally raised.

(*a*) It is said that the number of fossils in successive strata does not increase steadily as we ascend to modern times—that the numerical strength of the fauna is strangely irregular. Thus (in 1872) it was computed that 10,000 species were known from the early Silurian rocks, while the much later Permian yielded only 300. But those who use such arguments should mention that a large number of the Silurian species were discovered by the marvellous industry of one man in a favourable locality, and that the rocks of the Permian system are ill adapted for the preservation of fossils. Moreover, we cannot compute the relative duration of the different periods, we cannot infer evolutionary progress from the number of species, and we must make many allowances for the imperfections of the record.

(b) It is said that the occurrence of Fishes in the Silurian, and of many highly organised Invertebrates in the still earlier Cambrian, is inconsistent with a theory which would lead us to expect very simple fossil forms to begin with. But to say so is to forget that we have no conception of the vast duration of periods like the Silurian and Cambrian, while the antecedent Archæan rocks in which we might look for traces of simple ancestral organisms have been shattered and altered too thoroughly to reveal any important secrets as to the earliest animals.

(c) It is maintained that organic evolution proceeds very slowly, and that the geologists and biologists demand more millions than the experts in astronomical physics can grant them. But there is considerable difference of opinion as to the unthinkable length of time during which the earth may have been the home of life; we are apt to measure the rate of evolutionary change by the years of a man's lifetime which lasts but for a geological moment; and there is reason to believe that the simpler animals would change and take great steps of progress much more rapidly than those of high degree.

6. **Relative Antiquity of Animals.**—I have not much satisfaction in submitting the following table showing the relative antiquity of the higher animals. Such a table is only an approximation; it does not suggest the great differences in the duration of the various periods, nor how the classes of animals waxed and waned, nor how some types in these classes dropped off while others persisted. But the general fact which the table shows is true,—in the course of time higher and higher forms of life have come into being. It is true that the remains of mammals are of more ancient date than those of birds, but it is likely that the remains of the earliest birds have still escaped discovery; moreover, the earliest known mammalian remains seem to be of those of very simple types.

CHAP. XIII   *The Past History of Animals*   209

Primary or Palæozoic						Secondary or Mesozoic			Tertiary or Cainozoic			Quaternary or Post-Tertiary
Archæan	Cambrian	Silurian	Devonian	Carboniferous	Permian	Triassic	Jurassic	Cretaceous	Eocene	Miocene	Pliocene	
Many Invertebrates												
		Fishes										
				Amphibians								
						Reptiles						
								Birds				
								Mammals				
											Man?	

CHAPTER XIV

THE SIMPLEST ANIMALS

1. *The Simplest Forms of Life*—2. *Survey of Protozoa*—3. *The common Amœba*—4. *Structure of the Protozoa*—5. *Life of Protozoa*—6. *Psychical Life of the Protozoa*—7. *History of the Protozoa* 8. *Relation to the Earth*—9. *Relation to other Forms of Life*—10. *Relation to Man*

1. **The Simplest Forms of Life.**—It is likely that the first breath of life was in the water, for there most of the simplest animals and plants have their haunts. Simple they **are**, as an egg **is** simple when contrasted with a bird. They are (almost all) unit specks of living matter, each comparable to, but often more complex than, one of the numerous unit elements or cells which compose any higher plant or animal, moss or oak-tree, sponge or man. It is not merely because they are small that we cannot split them into separate parts different from one another,—size has little to do with complexity,—but rather because they are unit specks or single cells. But they are not "structureless"; in fact, old Ehrenberg, who described some of them in 1838 as "perfect organisms" and fancied he saw stomachs, vessels, hearts, and other organs within them, was nearer the truth than those who reduce the Protozoa to the level of white of egg.

Nor **are** they omnipresent, swarming in any drop of water. The clear water of daily use will generally disappoint, or rather please us by showing little trace of living things. But take a test-tube of water from a stagnant pool, hold it between your eyes and the light, and it is likely that you will see many forms of life. Simple plants and simple animals are there, the former represented by threads, ovals, and spheres in green, the latter by more mobile almost colourless specks or whitish motes which dance in the water. But besides these there are jerky swimmers whose appearance almost suggests their popular name of "water-fleas," and wriggling "worms," thinner than thread and lither than eels : both of these may be very small, but closer examination shows that they have parts and organs, that they are many-celled not single-celled animals.

Vary the observations by taking water in which hay stems or other parts of dusty dead plants have been steeped for a few days, and even with the unaided eye you will see a thick crowd of the mobile whitish motes which, from their frequent occurrence in such infusions, are usually called Infusorians. Or if a piece of flesh be allowed to rot in an open vessel of water, the fluid becomes cloudy and a thin flaky scum gathers on the surface. If a drop of this turbid liquid be examined with a high power of the microscope, you will see small colourless rods and spheres, quivering together or rapidly moving in almost incalculable numbers. These, though without green colour, are the minutest forms of plant life; they are Bacteria or Bacilli, the practically omnipresent microbes, some of which as disease germs thin our human population, while others as cleansers help to make the earth a habitable dwelling-place.

2. **Survey of Protozoa.**—Three great types of unicellular animals or Protozoa have been recognised in almost every classification.

(*a*) The Infusorians, so abundant in stagnant water, have a common character of activity expressed in the possession of actively mobile lashes of living matter known as cilia or flagella. Thus the slipper-animalcule (*Paramæcium*) is covered with rows of lashing cilia, while smaller, equally common forms, generally known as Monads, are borne along by the undulatory movement of one or two long whips or flagella. The bell-animalcules (*Vorticella*) which live in crowds,—a white fringe on the water weeds,—are generally fixed by stalks, but are crowned with active cilia at the upper end of the somewhat urn-shaped cell.

(*b*) In marked contrast to these are the parasitic Protozoa, the Gregarines, which infest most backboneless animals, notably the male reproductive organs of the earthworm or the gut of lobster and cockroach. Sluggishness and the absence of all locomotor processes are their characteristics.

(*c*) Between these two extremes of activity and passivity there is a third type well represented by the much-talked-of *Amœba* which glides about on the mud of the pond, by the sun-animalcules (*Actinosphærium*) which float in the clear water of brooks, by the limy-shelled, chalk-forming Foraminifera which move slowly on seaweeds or at the bottom of shallow water, or in some cases float at the surface of the sea, and by the flinty-shelled Radiolarians which live in the open ocean. In all these the living matter spreads out in thick or thin, stiff or plastic, free or interlacing processes, which often admit of a slow gliding motion, and are still more useful in surrounding minute food particles. To these root-like processes, which are capable of very considerable, often almost constant, change, these Protozoa owe their general name of Rhizopods. In contrast to the two preceding types which have definite boundaries or "skins," the Rhizopods are naked, and their living matter may overflow at any point.

FIG. 40.—A foraminifer (*Polystomella strigillata*) with interlacing processes of the living matter flowing out on all sides. Magnified 10 times. (From Chambers's *Encyclop.*; after Max Schultze.)

As the Infusorians are for the most part provided with cilia from which flagella differ only in detail, we may speak of the type as ciliated; the self-contained Gregarines, often wrapped up within a sheath, we may call predominantly encysted; while those forms which are intermediate between these two extremes, and exhibit outflowing processes of living matter, are called amœboid in reference to their most familiar type, the common Amœba.

But though the members of each class are characterised by the predominance of one of the three phases of cell-life, they sometimes pass from one phase to another. Thus the ciliated or the amœboid units may become encysted.

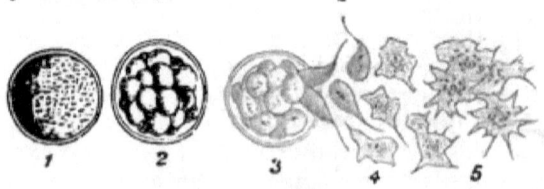

FIG. 41.—Protomyxa. 1, encysted; 2, dividing into many units; 3, these escaping as flagellate cells; 4, sinking into an amœboid phase; 5, fusing into a plasmodium. (From Chambers's *Encyclop.*; after Haeckel.)

As the three phases represent the three physiological possibilities of cell-life, it is natural to find that the very simplest Protozoa, such as *Protomyxa*, exhibit a cycle of amœboid, encysted, and flagellate phases, not having taken a decisive step along any one of the three great paths. Moreover, the cells of higher animals may be classified in the same way. The ciliated cells of the windpipe or the mobile spermatozoa correspond to Infusorians; mature ova, fat-cells, degenerate muscle-cells, correspond to Gregarines, while white blood-corpuscles and young ova are amœboid.

3. **The common Amœba.**—To find Amœbæ, which is not always easy, some water and mud from a pond should be allowed to settle in a glass vessel. Samples from the surface of the sediment should then be removed in a glass tube or pipette, dropped on a slide, and patiently examined under the microscope. Among the débris, traversed in most cases by swift Infusorians, the sought-for *Amœba* may be seen, as an irregular mass of living matter, often obscured with various kinds of particles and minute Algæ which it has engulfed, but hardly mistakable as it ploughs its way leisurely among the sediment, sending out blunt and changeful finger-like processes in the direction towards which it moves, and drawing in similar processes at the opposite side. From some objects it recoils, while others of an edible sort it surrounds with its blunt processes and gets outside of. Intense light makes it contract, and a minute drop of some obnoxious reagent causes it to round itself off and lie quiescent. Such is the simple animal which, in 1755, an early microscopist Rösel von Rosenhof was delighted to describe, calling it the "Proteus animalcule."

4. **Structure of the Protozoa.**—Most of these Protozoa are units or single cells, but this contrast between them and the higher animals is lessened by the fact that many Infusorians, some Radiolarians, and some of the very lowest forms live in close combination, a number of apparent individuals being substantially united in co-operation. In two quite different ways this compound life of some Protozoa arises. The "Flower of Tan" (*Fuligo* or *Æthalium septicum*) which in the summer months spreads as a yellowish slime on the bark of the tanyard, and supplies the student with the "largest available masses of undifferentiated protoplasm," arises from the flowing together and fusion of a number of smaller amœboid units. But in some Infusorians and Radiolarians the colony arises quite otherwise. Protozoa multiply by division; each unit splits into two which thenceforth live separate lives, and by and by themselves divide. Suppose, however, that the unit divide incompletely; suppose that the daughter-units, distinct though unsevered, redivide, and that the process is continued; a "colonial" Protozoon is the result. In this case the units do not flow together, they were never separated. But the "wisdom" of some of these

early associations has been justified in their far-off children, for in this way the many-celled animals began.

The cell-substance of a Protozoon is living matter, along with nutritive materials which are approaching that climax, and waste materials into which some of the cell substance has disintegrated. The cell has a kernel or nucleus, or more than one, essential to its complete life. There are bubbles of water taken in along with food particles, and in nearly all freshwater forms there are one or two special regions of internal activity, pulsating cavities or contractile vacuoles, which become large and small sometimes rhythmically, and may burst open on the surface of the cell. They are believed to help in getting rid of waste, and also in internal circulation. There is a rind in the Infusorians and Gregarines, and shells of flint and lime are characteristic of most Foraminifers and Radiolarians.

5. **Life of Protozoa.**—The life-histories of the Protozoa are very varied, but some chapters are common to most. They expend energy in movement; they regain this by feeding; their income exceeds their expenditure, and they grow; at the limit of growth they reproduce by dividing into two or many daughter-units; in certain states two individuals combine, either interchanging nuclear elements (in the ciliated Infusorians) or fusing together (as in some Rhizopods); in drought or in untoward conditions, or before manifold division, they often draw themselves together and encyst within a sweated-off sheath.

The Protozoa often multiply very rapidly. One divides into two, the two become four, and in rapid progression the numbers increase. On Maupas's calculation a single Infusorian may in four days have a progeny of a million. The same observer has shed a new light on another process—that of conjugation, the temporary or permanent union of two Protozoa, which in the ciliated Infusorians involves an interchange of nuclear particles. In November 1885, Maupas isolated an Infusorian (*Stylonichia*) and observed its generations till March 1886. By that time there had been two hundred and fifteen generations produced by ordinary division, but since these lowly organisms do not conjugate with near relatives, conjugation had not occurred. The result, corroborated in other cases, was striking. The whole family became exhausted, small, and "senile"; they ceased to divide or even to feed; their nuclei underwent a strange degeneration; they began to die. But individuals removed before the process had gone too far were observed to conjugate with unrelated forms and to live on. The inference was obvious. Conjugation in these Infusorians is of little moment to any two individuals; during long periods it need never occur, but it is essential to the continued life of the species. "It is a necessary condition of their eternal youth."

We must return, however, to the everyday life of the Protozoa. Rhizopods move by means of outflowing processes of their living matter which stream out at one corner and are drawn in at another; the Infusorians move more rapidly by undulating flagella or by numerous cilia which work like flexible oars; the parasitic Gregarines without any definite locomotor structures sometimes writhe sluggishly. A few Infusorians have a spasmodic leaping or springing motion, while the activity of others (like *Vorticella*) which in adult life are fixed, is restricted to the contraction and expansion of a stalk and to the action of cilia around the opening which serves as a mouth. *Arcella* is aided in its movements by the formation of gas bubbles in different parts of its cell-substance.

The food consists of other Protozoa, of minute Algæ, and of organic débris, simply engulfed by the Amœbæ, wafted by cilia into the "mouth" of most Infusorians. The parasitic Gregarines absorb the débris of the cells or tissues of the animals in which they live, while not a few suck the cell-contents of freshwater Algæ like *Spirogyra*. A few Protozoa are green, and some are able to use carbonic acid after the manner of plants. Almost all Radiolarians and a few Foraminifers live in constant and mutually helpful partnership or symbiosis with small Algæ which flourish within their cell-substance.

As to the other functions, the cells absorb oxygen and liberate carbonic acid, digest the food-particles and excrete waste, produce cysts or elaborate shells.

6. **Psychical Life of the Protozoa.**—We linger over the Protozoa because they illumine the beginnings of many activities, and we cannot leave them without asking what light they cast upon the conscious life of higher animals. Is the future quite hidden in these simple organisms or are there hints of it?

According to some, the Protozoa, with frequently rapid and useful movements, with capacities for finding food and avoiding danger, with beautiful and intricate shells, are endowed with the will and intelligence of higher forms of life. According to others, their motions are arbitrary and without choice, they are only much more complex than those of the potassium ball which darts about on the surface of water, the organisms are drawn by their food instead of finding it, their powers of selection are sublimed chemical affinities, their protective cysts are quite necessary results of partial death, and their houses are but crystallisations. In both interpretations there is some truth, but the first credits the Protozoa with too much, the second with too little.

Cienkowski marvelled over the way in which *Vampyrella* sought and found a *Spirogyra* filament and proceeded to suck its contents; Engelmann emphasised the wonderful power of adjustment in *Arcella* which evolves gas bubbles and thus rises or rights itself when capsized, and also detected perception and decision in the motions of young *Vorticella* or in the pursuit of one unit by another; Oscar Schmidt granted them only "a very dim general feeling" and the power of responding in different ways to definite stimuli; Schneider believed that they acted on impulses based upon definite impressions of contact; Moebius would credit them with the power of reminiscence and Eimer with will.

Romanes finds evidence of the power of discriminative selection among the protoplasmic organisms, and he quotes in illustration Dr. Carpenter's account of the making of shells. "Certain minute particles of living jelly, having no visible differentiation of organs . . . build up 'tests' or casings of the most regular geometrical symmetry of form and of the most artificial construction. . . . From the same sandy bottom one species picks up the coarser quartz grains, cements them together with phosphate of iron (?) which must be secreted from their own substance, and thus constructs a flask-shaped 'test' having a short neck and a single large orifice. Another picks up the finer grains and puts them together with the same cement into perfectly spherical 'tests' of the most extraordinary finish, perforated by numerous small tubes, disposed at pretty regular intervals. Another selects the minutest sand-grains and the terminal points of sponge spicules, and works these up together apparently with no cement at all, but by the 'laying' of the spicules into perfect spheres, like homœopathic globules, each having a single fissured orifice." This selecting power is marvellous; we cannot explain it; the animals are alive and they behave thus. But it must be remembered that even 'dead' substances have attractive affinities for some things in preference to others, that the cells of roots and those lining the food-canal of an animal or floating in its blood show a power of selection. Moreover, if we begin with a unit which provides itself with a coating of sponge spicules, at first perhaps because they were handiest, it is not difficult to understand why the future generations of that species should continue to gather these minute needles. Being simply separated parts of their parents, whose living matter had become accustomed to the stimulus of sponge spicules, the descendants naturally sustain the tradition. This organic memory all Protozoa must have, for the young are separated parts of the parents.

Haeckel was one of the first (1876) to urge the necessity of recognising the "soul" of the cell. He maintained that the continuity of organic life led one to assume a similar continuity of psychical life, that an egg-cell had in it not only the potency of forming tissues and organs but the rudiments of a higher life as well, that the Protozoa likewise must be regarded not only as physical but as psychical, in fact that the two are inseparable aspects of one reality. "The cell-soul in the monistic sense is the sum-total of the energies embodied in the protoplasm, and is as inseparable from the cell-substance as the human soul from the nervous system." For several years Verworn has been investigating the psychical life of the Protozoa. He has conducted his researches with great care and thoroughness, observing the animals both in their natural life and in artificial conditions. I shall cite his conclusions, translating them freely: "An investigator of the psychical processes in Protists (simple forms of life) has to face two distinct problems. The first is comparative, and inquires into the grade of psychical development which the Protists may exhibit—the known standard being found of course in man; the second is physiological, and inquires into the nature of these psychical processes. Since we know these only through the movements in which they are expressed, the investigation is primarily a study of the movements of Protists.

"On a superficial observation of these movements the impression arises in the observer's mind that they are the result of higher psychical processes, like the consciously willed activities of men. Especially the spontaneous movements of advance and recoil, of testing and searching, give us the impression of being intentional and voluntary, since no external stimulus can account for them; while even some of the movements provoked by stimuli appear on account of their marked aptness to arise from conscious sensation and determination.

"But a critical study of the results yielded by an investigation of spontaneous and stimulated movements warrants a more secure judgment than that of the superficial observer, and leads to a conclusion opposed to his. To this conclusion we are led, that none of the higher psychical processes, such as conscious sensations, representations, thoughts, determinations, or conscious acts of will, are exhibited by Protists. A number of criteria show that the movements are in part impulsive and automatic, and in part reflex, and in both cases expressions of unconscious psychical processes.

"This opinion is corroborated by an examination of the structure of these Protists, for this does not seem such as would make it possible for the individual to have an idea of its own unified self, and the absence of self-consciousness excludes the higher psychical processes. Small fragments cut from a Protist cell continue to make the same movements as they made while parts of the intact organism. Each fragment is an independent centre for itself. There is no evidence that the nucleus of the organism is a psychical centre. There is no unified Psyche.

"Since the characteristic movements persist in such small fragments, they cannot be the expression of any individual consciousness, for the individuality has been cut in pieces."

The dilemma is obvious; either there are no psychical processes in the Protists, or they are inseparable from the molecular changes which occur in the parts of the material substance.

If no psychical processes occur in the Protists, where do they begin? There is no distinct point in the animal series at which a nervous system may be said to make its first appearance. If there are none, even rudimentarily, in the Protists, then these simple organisms do not potentially include the life of higher organisms. If there are none in the Protists, are there any in the germs from which men develop?

Verworn seizes the other horn of the dilemma, maintaining that the superficial observers are wrong in crediting the Protozoa with their own intelligence or with some of it, but right in concluding that psychical processes of some sort are there. But since he cannot in any way locate these processes, since he finds that even small fragments retain their life for a time and behave much as the entire cells did, he maintains that all life is psychical.

7. **History of the Protozoa.**—We know that the Protozoa have lived on the earth for untold ages, for the shells of Foraminifera and others may be disentombed from almost the oldest rocks. The word Protozoa, a translation of the German *Urthiere* or primitive animals, suggests that the Protozoa are not only the simplest, but the first animals, or the unprogressive descendants of these. Nowadays we can hardly feign to consider this proposition startling, for we know that all the higher animals, including ourselves, begin life at the beginning again as single cells. From the division and redivision of an apparently simple fertilised egg-cell an embryo is built up which grows from stage to stage till it is hatched, let us say, as a chick. It is only necessary to extend this to the wider history of the race. What the egg is to the chick the original Protozoa were to the animal series; the present Protozoa are like eggs which have lived on as such without making much progress.

We do not know how the Protozoa began to be upon the earth, whether they originated from not living matter or in some yet more mysterious way. The German naturalist Oken, a prominent type of the school of "Natural Philosophers" who flourished about the beginning of this century, dreamed of a primitive living slime (*Urschleim*) which arose in the sea from inorganic material. His dream was prophetic of the modern discovery of very simple forms of life, in connection with one of which there is an interesting and instructive story. That one, perhaps I should say that supposed one, was called *Bathybius*, and since those who are eager to make

The Simplest Animals

points against science (that is to say against knowledge) always tell the story wrongly, I shall make a digression to tell it rightly.

In 1857 Captain Dayman, in charge of a vessel engaged in connection with cable-laying, discovered on the submarine Atlantic plateau the abundant presence of slimy material which looked as if it were alive. Preserved portions of this formless slime were afterwards described by Huxley, and he named the supposed organism, partly from its habitat, partly after his friend Haeckel, *Bathybius Haeckelii*. On the *Porcupine* expedition Professors Wyville Thomson and Carpenter observed it in its fresh state, and Haeckel afterwards described some preserved specimens. Its interest lay in its simplicity and apparent abundance; Oken's dream seemed to be coming true; it seemed as if life were a-making in the still depths.

But when the *Challenger* expedition went forth, and the bed of the ocean was explored for the first time carefully, the organism *Bathybius* was nowhere to be found. But this was not all; the cruellest blow was yet to come. Dr. John Murray saw reason to suspect that Bathybius was not an organism at all, that it could be made in a test-tube, and was nothing but a gelatinous form of sulphate of lime precipitated from the sea water by the action of the alcohol in the preserving vessels. He renounced Bathybius, Wyville Thomson acquiesced, Huxley surrendered his organism to the chemists, and the obscurantists rejoiced exceedingly over the mare's nest. Bathybius became famous, it was trotted out to illustrate the fallibility of science, a useful if it were not a somewhat superfluous service.

But the non-existence of Bathybius was not proved by the fact that the *Challenger* explorers failed to find it, nor was it certain that Murray's destructive criticism covered all the facts. Haeckel clung with characteristic pertinacity to Bathybius, and his constancy has been to some extent justified by the fact that in 1875 Bessels, on a North Polar expedition, dredged from 92 fathoms of water in Smith's Sound abundant quantities of a closely-similar slime. He observed its vital movements, and called it *Proto-Bathybius*. It may be that it consists of the broken-off portions of Foraminifera; we require to know yet more about it, but I have said enough to show that it is unfair to stop telling the story with the words "mare's nest." But whether there be a Bathybius, a Proto-Bathybius, or no Bathybius at all, we are as students of science compelled to confess our complete ignorance as to the origin of life.

8. **Relation to the Earth.**—The floor of the sea for a variable number of miles (not exceeding 300) from the shore is covered with a heterogeneous deposit, washed in great part from the nearest continent. In this deposit shells of Foraminifera usually

occur, but they become more numerous farther from the land, where the floor of the sea is often covered with a whitish "ooze," most of which consists of Foraminifera which in dying have sunk from the surface to the bottom. They are forming the chalk of a possible future, just as many chalk-cliffs and pure limestones represent the ooze of a distant past. In other regions the hard parts of Radiolarians or Diatoms (small plants) or Pteropods (minute molluscs) are very abundant. As the Foraminifers have made much of the chalk, so Radiolarians have formed less important siliceous deposits, such as the Barbados Earth, from which Ehrenberg described no fewer than 278 species. At marine depths greater than 2500 fathoms the Globigerina or other Foraminifer shells are no longer present, not because there are none at the surface, but apparently owing to the solution of the shells before they reach such a vast depth. Here the floor is covered with a very fine **reddish or brownish deposit**, often called "red-clay," a very **heterogeneous deposit of meteoric and** volcanic **dust and of residues of surface-animals.** Along with this, in some of the very deepest parts, *e.g.* of the Central Pacific, **there are** accumulations of Radiolarian shells, which do not readily dissolve.[1]

9. **Relation to other Forms of Life.**—On the one hand the Protozoa are devourers of organic débris and the enemies of many small plants; on the other hand they form the fundamental food of higher animals, helping, for instance, to make that thin sea-soup on which many depend. Moreover, among them there are many parasites both on vegetable and animal hosts.

10. **Relation to Man.**—In many indirect **ways** these firstlings affect human life, nor are there wanting direct points of contact; witness a few Protozoa parasites in man, an Amœba, some Gregarines, and some Infusorians, which are very trivial, however, in comparison with the numerous plant-parasites—the Bacteria.

Among the earliest human records of Protozoa is the notice which Herodotus and Strabo **take of the large** coin-like Nummulites, the "Pharaoh's beans" **of popular** fancy. But the minuteness of most Protozoa kept them out of sight for ages. They were virtually discovered by Leeuwenhoek (b. 1632) about the middle of the seventeenth century, and soon afterwards demonstrated by Hooke to the Royal Society of London, the members of which signed an affidavit that they had really seen them! In 1755 Rösel von Rosenhof discovered the Amœba, or "Proteus animalcule;" but his discovery was ineffective till Dujardin in 1835 demonstrated the simplicity of the Foraminifers, and till Von Siebold in 1845

---

[1] For details, see conveniently H. R. Mill's *Realm of Nature* (Lond. 1892).

showed that Infusoria were single cells comparable to those which make up a higher animal. For the resemblance between some of the spirally twisted shells of Foraminifera and those of the immensely larger Molluscan Ammonites and Nautili led many to maintain that the Foraminifera were minute predecessors or else dwindling dwarfs of the Ammonites. So Ehrenberg (1838) figured the presence of many organs within the Infusorian cell. But as the microscope was perfected naturalists were soon convinced that the Protozoa were unit masses of living matter. This is their great interest to us; they are, as it were, higher organisms analysed into their component elements. We see them passing through cycles of phases, from ciliated to amœboid, from amœboid to encysted, cycles which shed light upon changes both of health and of disease in higher animals. Again, they seem like ova and spermatozoa which have never got on any farther.

CHAPTER XV

BACKBONELESS ANIMALS

1. *Sponges*—2. *Stinging-Animals or Cœlenterata*—3. "*Worms*"—
4. *Echinoderms*—5. *Arthropods*—6. *Molluscs*

1. **Sponges.**—Sponges are many-celled animals without organs, with little division of labour among their cells. A true "body" is only beginning among sponges.

Adult sponges are sedentary, and plant-like in their growth. With the exception of the freshwater sponge (*Spongilla*) they live in the sea fixed to the rocks, to seaweeds and to animals, or to the muddy bottom at slight or at great depths. They feed on microscopic organisms and particles, borne in with currents of water which continually flow through the sponge. The sponge is a Venice-like city of cells, penetrated by canals, in which incoming and outflowing currents are kept up by the lashing activity of internal ciliated cells. These ciliated cells, on which the whole life of the sponge depends, line the canals, but are especially developed in little clusters or ciliated chambers. The currents are drawn in through very small pores all over the surface; they usually flow out through much larger crater-like openings.

Sponges feed easily and well, and many of them grow out in buds and branches. A form which was at first a simple cup may grow into a broad disc or into a tree-like system. And as trees are blown out of shape by the wind, so sponges are influenced by the currents which play around them, as well as by the nature of the objects on which they are fixed. Like many other passive organisms, sponges almost always have a well-developed skeleton, made of flinty needles and threads, of spicules of lime, or of fibres of horn-like stuff. While sponges do not rise high in organic rank, they have many internal complications and much beauty.

Sponges may be classified according to their skeleton, as

calcareous, flinty, and horny. (*a*) The calcareous forms with needles of lime have a world-wide distribution in the sea, from between tide-marks to depths of 300 to 400 fathoms. They often retain a cup-like form, but vary greatly in the complexity of their canals. The sac-like *Sycandra* (or *Grantia*) *compressa* is common on British shores. (*b*) The siliceous sponges are more numerous, diverse, and complicated, and the flinty needles or threads are often combined with a fibrous "horny" skeleton. Venus'-Flower-Basket (*Euplectella*) has a glassy skeleton of great beauty, Mermaids' Gloves (*Chalina oculata*) with needles of flint and horny fibres is often thrown up on the beach, the Crumb-of-Bread Sponge (*Halichondria panicea*) spreads over the low-tide rocks. Some have strange habits, witness *Clione* which bores holes in oyster shells, or *Suberites domuncula* which clothes the outside of a whelk or buckie shell tenanted by a hermit-Crab. Unique in habitat is the freshwater sponge (*Spongilla*) common in some canals and lakes, notable for plant-like greenness, and for the vicissitudes of its life-history. (*c*) The "horny" sponges which have a fibrous skeleton but no proper spicules are well represented by the bath-sponges (*Euspongia*) which thrive well off Mediterranean coasts, where they are farmed and even bedded out.

Sponges are ancient but unprogressive animals. Their sedentary habits, from which only the embryos for a short time escape, have been fatal to further progress. They show tissues as it were in the making. They are living thickets in which many small animals play hide-and-seek. Burrowing worms often do them much harm, but from many enemies they are protected by their skeletons and by their bad taste.

2. **Stinging-Animals or Cœlenterata.**—It is difficult to find a convenient name for the jellyfish and zoophytes, sea-anemones and corals, and many other beautiful animals which are called Cœlenterates; but the fact that almost all have poisonous stinging lassoes in some of their skin-cells suggests that which we now use.

Representatives of the chief divisions may be sometimes found in a pool by the shore. Ruddy sea-anemones, which some call sea-roses, nestle in the nooks of the rocks; floating in the pool and throbbing gently is a jellyfish left by the tide; fringing the rocks are various zoophytes, or, if we construe the name backwards plant-like animals; besides these, and hardly visible in the clear water, are minute translucent bells some of which have a strange relationship with zoophytes; and there are yet other exquisitely delicate, slightly iridescent globes—the Ctenophores which move by comb-like fringes of cilia. But we must search an inland pool to find one of the very simplest members of this class—the freshwater *Hydra* which hangs from the floating duckweed and other plants.

This *Hydra* is a tubular animal often about quarter of an inch in length. One end of the tube is fixed, the other bears the mouth surrounded by a crown of mobile tentacles. It is so simple that cut-off fragments if not too minute may grow into complete animals; when well fed, the *Hydra* buds out little polypes like itself, and these are eventually set free.

If we suppose the budding of *Hydra* continued a hundred-fold, till a branched colony of connected individuals is formed, we have an idea of a hydroid or zoophyte colony. For a zoophyte is a colony of many hydra-like polypes, which are supported by a continuous outer framework and share a common life. Numerous as may be the "persons" on a branched hydroid, all have arisen from one more or less Hydra-like individual.

Sometimes, however, there is a marked division of labour in such a colony, as in *Hydractinia* which has nutritive, reproductive, sensitive, and perhaps also protective "persons,"—three or four castes into which the colony is divided. The difference between nutritive and reproductive members is often well marked, and this has a special interest in the case of many zoophytes. For many of these, especially among those known as Tubularians and Campanularians, have reproductive individuals which are set adrift as small swimming-bells or medusoids, somewhat like miniature jellyfish. A fixed plant-like, asexual hydroid colony buds off free-swimming, sexual medusoids, from the fertilised eggs of which embryos develop which grow into hydroids. This is known as alternation of generations, and is a remarkable illustration of activity and passivity combined in one life-cycle.

But all the miniature jellyfish in the sea are not the liberated reproductive buds of hydroid colonies. Some which are in structure exceedingly like the liberated medusoids never have any connection with a hydroid. Their embryos grow into medusoids like the parents. Quite distinct from these medusoids, though sometimes superficially like them, are the true jellyfishes which are sometimes stranded in great numbers on the beach. These medusæ belong to a different series, and some of their features link them rather to the sea-anemones than to the hydroids.

The sea-anemones and the corals are tubular animals whose mouths are encircled by tentacles, but they are more complicated internally than the polypes of the *Hydra* or hydroid type. For the latter are simple tubes, while the sea-anemones and their relatives have turned-in lips which make a kind of gullet, and the inside tube thus formed is connected with the outer wall of the body by many radiating partitions some idea of which can be gained by looking at the skeletons or shells of many corals. Related to the sea-anemones but different in some details, are many colonies, of which Dead-men's-fingers (*Alcyonium digitatum*) is a common type. Animals resembling sea-anemones have often much lime about them, and the same is true of others which resemble *Alcyonium*; in both cases we call these calcareous forms corals.

In this bird's-eye view of Stinging-animals, we have recognised the great types, but we have left others of minor importance out of account, especially certain corals belonging to the hydroid series and known as Millepores, also the Portuguese Man-of-War and its relatives (Siphonophora), which are colonies of more or less medusoid individuals with much division of labour, and lastly the Ctenophores, such as *Beroë* and *Pleurobrachia*, which represent the climax of activity among Cœlenterates.

A brief recapitulation will be useful :

First Series—Hydroid and Medusoid types (Hydrozoa):—

(1) The freshwater *Hydra* and a few forms like it.
(2) The hydroids or zoophytes, each of which may be regarded as a compound much-branched Hydra; including (*a*) many whose reproductive persons are not liberated, especially Sertularians and Plumularians;—(*b*) many whose reproductive persons are liberated as swimming bells or medusoids, especially Tubularians and Campanularians.
(3) Free medusoids, anatomically like the liberated bells of 2 (*b*), but without any connection with zoophytes.
(4) A few colonial medusoids such as the Portuguese Man-of-War (*Physalia*).
(5) A few hydroid corals or Millepores.

Second Series—Jellyfish and Sea-Anemone types (Scyphozoa):—

(1) The true jellyfishes or Medusæ, including (*a*) a form like *Pelagia* which is free-swimming all its life through, (*b*) the common *Aurelia* whose embryos settle down and become polypes from which the future free-swimming jellyfishes are budded off, (*c*) the more or less sedentary jellyfish known as Lucernarians.
(2) The sea-anemones and their relatives, including (*a*) sea-anemones proper (*e.g. Actinia*) and their related reef-building coral-colonies (*e.g.* star-corals *Astræa*, brain-coral *Mæandrina*); and (*b*) Dead-men's-fingers (*Alcyonium*) and others like it, also with related corals, *e.g.* the organ-pipe coral (*Tubipora musica*) and the "noble coral" of commerce (*Corallium rubrum*).

Third Series—

The Ctenophores, which are markedly contrasted with corals, being free and light and active. Many (*e.g. Beroë* and *Pleurobrachia*) swarm in our seas in summer, iridescent in daylight, phosphorescent at night. They differ in many ways from other Cœlenterates, thus the characteristic stinging cells are modified into adhesive cells.

The first and second series, separated by differences of structure and development, are yet parallel. In both there are polype-types; in both medusoid types; in both there are single individuals and colonies of individuals; in both there are "corals." We may compare a *Hydra* with a sea-anemone, a medusoid with a jellyfish, a hydroid colony with Dead-men's-fingers, Millepores with

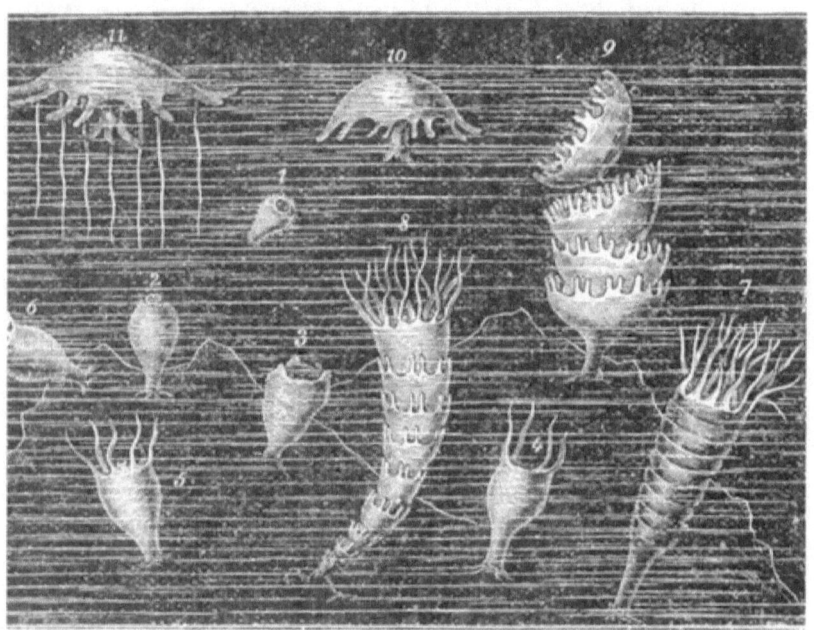

FIG. 42.—The alternation of generations in the common jellyfish *Aurelia*. 1, the free-swimming embryo; 2, the embryo settled down; 3, 4, 5, 6, the developing asexual stages, or hydra-tubæ; 7, 8, the formation of a pile of individuals by transverse budding; 9, the liberation of these individuals; 10, 11, their progress towards the free-swimming sexual medusa form. (From the *Evolution of Sex*; after Haeckel.)

the commoner reef-corals. Moreover, we may compare a medusoid liberated from a hydroid with *Aurelia* liberated from its fixed polype-stage, and permanently-free medusoids with jellyfishes like *Pelagia*. These are physiological parallels.

The sedentary polypes are somewhat sluggish, with a tendency to bud and to form shells or skeletons of some kind. The free-swimming medusoid types are active, they rarely bud, they do not form skeletons, but their activity is sometimes expressed in phosphorescence, and their fuller life is associated with the development of sense-organs and a more compacted nervous system. In both sets the food usually consists of small organisms, in securing which the tentacles and the stinging cells are of use.

All the Stinging-animals are marine except the species of *Hydra*, a minute relative called *Microhydra*, the hydroid *Cordylophora* which occurs in brackish water and in canals, a strange form *Polypodium* which is parasitic in its youth on the eggs of the Russian sturgeon or sterlet, and a freshwater jellyfish (*Limnocodium*) which was found in the tanks at Kew. The rest live in the sea. Hydroids grow on rocks and shells and on the backs of crabs and other animals which they mask; sea-anemones live on the shore-rocks —but not a few are found at considerable depths; the medusoid types frequent the opener sea where Siphonophores and Ctenophores bear them company.

Various kinds of corals should be contrasted. Dead-men's-fingers with numerous jagged spicules of lime in its flesh is just beginning to be coralline. Similar spicules have been fused together in an external tube in the organ-pipe coral. In the red coral the calcareous material forms an axis around which the individuals are clustered. Very different are the reef-building corals, where the cup in which each individual lived is more or less well marked according as it has remained distinct or fused with its neighbours, and where an image of the fleshy partitions of the sea-anemone-like animal is seen in the radiating septa of lime.

Corals are passive, and like many animals of similar habit have calcareous shells, but how do they get the carbonate of lime of which these are composed? Is that salt—by no means abundant in sea-water—plentiful near coral-reefs, or is there a double-decomposition between the abundant calcium sulphate and the coral's waste-products, as has been suggested by Irvine and Murray? On what do the corals feed, for they seem always to be empty? Do their bright pigments enable them, as Hickson suggests, to feed like plants on carbonic acid?

The struggle for standing-room should also be thought of, and the throngs of gaily-coloured animals which browse and hide on the coral banks.

Many of the Stinging-animals have forms and colours which delight our eyes, and the quaint partnerships between sea-anemones and crabs are interesting.

But it is through corals that Cœlenterates come into closest touch with human life. For the stinging of bathers by jellyfish is a minor matter, and the thousands which are cast upon the beach are of no use as manure, being little more than animated sea-water.

As sponges showed tissues in the making, so among Stinging-animals organs begin—eyes and ears, nerve-rings, and special reproductive structures. The zoologist has much to learn in regard to the alternation of hydroid and medusoid in one life-cycle, the division of labour in *Hydractinia* and other colonies, and the meaning and making of a skeleton. Nor can we forget the long past in which there were ancient coral reefs, and types of coral hardly represented now, and strange Graptolites whose nature we do not yet clearly understand.

We begin the series of many-celled animals with Sponges and Cœlenterates, partly because they are on the whole simplest, but more precisely because their types of structure are least removed from that two-layered sac-like embryo or gastrula which recurs in the life-history of most animals, and which we have much warrant for regarding as a hint of what the first successful many-celled animals were like. The Sponges and Cœlenterates differ from the higher animals: (1) In retaining the symmetry of this gastrula, in being like it radially symmetrical, and in so growing that the axis extending from the mouth to the opposite pole corresponds to the long axis of the embryo; (2) in being two-layered animals, for between the outer skin and the lining of the internal food-cavity there is only a more or less indefinite jelly instead of a definite stratum of cells; (3) in having only one internal cavity, instead of having, like most other animals, a body-cavity within which a distinct food-cavity lies.

3. "**Worms**."—This title is one of convenience, without strict justification. For there is no class of "worms," but an assemblage of classes which have little in common. "Worm" is little more than a name for a shape, most of the animals so called differing from anemones and jellyfish in having head, tail, and sides. The simplest worms were apparently the first many-celled animals to move persistently head foremost, thus acquiring distinct bilateral symmetry, and a definite nervous centre or brain in that region which had most experience—the head. In our survey we are helped a little by the fact that many consist of a series of rings or segments, while others are all one piece or unsegmented. It is generally true that the latter are in structure simpler and more primitive than the former.

1st Set of Worms. **Plathelminthes or flat worms.** 1st Class.—Turbellaria or Planarians.—These are small worms, living in the sea or in fresh water, or occasionally in damp earth, covered externally with cilia, very simple in structure, usually feeding on minute animals. The genus *Planaria*, common in fresh water; green species of *Vortex* and *Convoluta*, which are said by some to owe their colour to minute partner algæ; *Microstoma*, which by budding forms temporary chains of eight or sixteen individuals as if suggesting how a ringed worm might arise; *Gunda*, with a hint of internal segmentation; and two parasitic genera—*Graffilla* and *Anoplodium*—may be mentioned as representatives of this class. You will find specimens by collecting the waterweeds from a pond or seaweeds from a shore-pool, and the simplicity of some may be demonstrated by observing that when they are cut in two each half lives and grows.

2nd Class.—Trematoda or Flukes. These are parasitic "worms," living outside or inside other animals, often flat or leaf-like in form, provided with adhesive and absorbing suckers. Those which live as ectoparasites, *e.g.* on the skin of fishes, have usually a simple history; while those which are internal boarders have an intricate life-cycle, requiring to pass from one host to another of a different kind if their development is to be fulfilled. Thus the liver-fluke (*Distomum hepaticum*), which causes the disease of liver-rot in sheep, and sometimes destroys a million in one year in Britain alone, has an eventful history. From the bile-ducts of the sheep the embryos pass by the food-canal to the exterior. If they reach a pool of water they develop, quit their egg-shells, and become for a few hours free-swimming. They knock against many things, but when they come in contact with a small water-snail (*Lymnæus truncatulus*) they fasten to it, bore their way in, and, losing their locomotor cilia, encyst themselves. They grow and multiply in a somewhat asexual way. Cells within the body of the encysted embryo give rise to a second generation quite different in form. The second generation similarly produces a third, and so on. Finally, a generation of little tailed flukes arises; these leave the water-snail, leave the water too, settle on blades of grass, and lose their tails. If they be eaten by a sheep they develop into adult sexual flukes. Others have not less eventful life-cycles, but that of the liver-fluke is most thoroughly known. If you dissect a frog you are likely to find *Polystomum integerrimum* in the lungs or bladder; it begins as a parasite of the tadpole, and takes two or three years to become mature in the frog. Quaint are the little forms known as *Diporpa* which fasten on the gills of minnows, and unite in pairs for life, forming double animals (*Diplozoon*); and hardly less strange is *Gyrodactylus*, another parasite on freshwater fishes, for three generations are often found together, one within the other. The most formidable fluke-parasite of man is *Bilharzia*, or *Distomum hæmatobium*, common in Africa.

3rd Class. Cestoda or Tapeworms. These are all internal parasites, and, with the exception of one (*Archigetes*), which fulfils its life in the little river-worm *Tubifex*, the adults always occur in the food-canal of backboned animals. Like the flukes, they have adhesive suckers, and sometimes hooks as well; unlike flukes and planarians, which have a food-canal, they absorb the juices of their hosts through their skins, and have no mouth or gut. Like the endo-parasitic flukes, the tapeworms have (except *Archigetes*) intricate life-histories. Both Turbellarians and Trematodes are small, rarely more than an inch at most in length, but the tapeworms may measure several feet. In the adult *Tænia solium*, which is sometimes found in the intestines of man, we see a small head like that of a pin; it is fixed by hooks and suckers to the wall of the food-canal; it buds off a long chain of "joints," each of which is complete in itself. As these joints are pushed by continued budding farther and farther from the head, they become larger, and distended with eggs, and even with embryos, for the bisexual tapeworm seems able to fertilise itself, which is a very rare thing among animals. The terminal joints of the chain are set free, one or a few at a time, and they pass down the food-canal to the exterior. The tiny embryos which they contain when fully ripe are encased in firm shells. It may be that some of them are eaten by a pig, the shells are dissolved away in the food-canal, small six-hooked embryos emerge. These bore their way into the muscles of the pig and lie dormant, increasing in size however, becoming little bladders, and forming a tiny head. They are called bladder-worms, and it was not till about the middle of this century that they were recognised as the young stages of the tapeworm. For if the diseased pig be killed and its flesh eaten (especially if half-cooked) by man, then each bladder-worm may become an adult sexual tapeworm. The bladder part is of no importance, but the head fixes itself and buds off a chain. For many others the story is similar; the bladder-worm of the ox becomes another tapeworm (*Tænia saginata*) in man; the bladder-worm of the pike or turbot becomes another (*Bothriocephalus latus*); the bladder-worm of the rabbit becomes one of the tapeworms of the dog, that of the mouse passes to the cat, and so on. A bladder-worm which forms many heads destroys the brain of sheep, etc., and has its tapeworm stage (*Tænia cœnurus*) in dog or wolf. Another huge bladder-worm, which has also many heads, and sometimes kills men, has also its tapeworm stage (*Tænia echinococcus*) in the dog. But enough of these vicious cycles.

2nd Set of Worms. **Ribbon Worms** or **Nemerteans**—4th Class, Nemertea.—In pleasing contrast to the flukes and tapeworms, the Nemerteans are free-living "worms." They are mostly marine, often brightly coloured, almost always elongated, always covered with cilia. There is a distinct food-canal with a posterior opening, a blood-vascular system for the first time, a well-developed nervous system, a remarkable protrusible "proboscis" lying in a sheath along the back, a pair of enigmatical ciliated pits on the head. The sexes are almost always separate. Almost all Nemerteans are carnivorous, but two or three haunt other animals in a manner which leads one to suspect some parasitism; thus *Malacobdella* lives within the shells of bivalve molluscs. We find many of them under loose stones by the sea-shore; one beautiful form, *Lineus marinus*, sometimes measures over twelve feet in length. Some, such as *Cerebratulus*, break very readily into parts, even on slight provocation, and these parts are said to be able to regrow the whole. To speculative zoologists, the Nemerteans are of great interest on account of the vertebrate affinities which some of their structures suggest. Thus the sheath of the "proboscis" has been compared with the vertebrate notochord (the structure which precedes and is replaced by a backbone), and the two ciliated head-pits with gill-slits.

3rd Set of Worms. **Nemathelminthes** or **Round-Worms**—5th Class, Nematoda or Thread-Worms.—The "worms" of this class are usually long and cylindrical, and the small ones are like threads. The skin is firm, the body is muscular; in most a simple food-canal extends from end to end of the body-cavity now for the first time distinct; the sexes are separate. Many of the Nematodes live in damp earth and in rottenness; many are, during part of their life, parasitic in animals or plants. We have already noticed how long some of them— "paste-eels," "vinegar-eels," etc.—may lie in a dried-up state without dying. The life-histories are often full of vicissitudes; thus the mildew-worm (*Tylenchus tritici*) passes from the earth into the ears of wheat, and many others make a similar change; the female of *Sphærularia bombi* migrates from damp earth into humble-bees, and there produces young which find their way out; others, *e.g.* some of the thread-worms found in man (*Oxyuris*, *Trichocephalus*), pass from water into their hosts; others are transferred from one host to another; as in the case of the *Trichina* with which pigs are infected by eating rats, and men infected by eating diseased pigs, or the small *Filaria sanguinis hominis*, sometimes found in the blood of man, which seems to pass its youth in a mosquito. Somewhat different from the other Nematodes are those of which the horse-hair worm *Gordius* is a type. They are sometimes found inside animals (water-insects, molluscs, fish, frog, etc.), at other times they appear in great numbers in the pools, being, according to popular superstition, vivified horse-hairs.

6th Class, Acanthocephala.—Including one peculiar genus of parasites (*Echinorhynchus*).

4th Series of Worms. **The Annelids** or **Ringed Worms** —7th Class, Chætopoda or Bristle-footed "worms."—In the earthworms (*Lumbricus*, etc.), in the freshwater worms (*Nais*, *Tubifex*, etc.), in the lobworms (*Arenicola piscatorum*), and in the sea-worms (*Nereis*, *Aphrodite*, etc.), all of which are ranked as Chætopods, the body is divided into a series of similar rings or segments, and there are always some, and often very many, bristles on the outer surface. The segments are not mere external rings, but divisions of the body often partially partitioned off internally, and there is usually some repetition of internal organs. Thus in each segment there are often two little kidney-tubes or nephridia, while reproductive organs may occur in segment after segment. Moreover, there are often two feet on each ring. The nervous system consists of a dorsal brain and of a double nerve-cord lying along the ventral surface. The nerve-cord has in each segment a pair of nerve-centres or ganglia, and divides in the head region to form a ring round the gullet united with the brain above. The existence of nerve-centres for each segment makes each ring to some extent independent, but the brain rules all. This type of nervous system represents a great step of progress; it is very different from that of Stinging-animals, which lies diffusely in the skin or forms a ring around the circumference; different from that of the lower "worms," where the nerve-cords from the brain usually run along the sides of the body; different from that of molluscs, where the nerve-centres are fewer and tend to be concentrated in the head; different finally from the central nervous system of backboned animals, for that is wholly dorsal. But the type characteristic of ringed "worms"—a dorsal brain and a ventral chain of ganglia—is also characteristic of crustaceans, insects, and related forms.

Of bristle-footed "worms," there are two great sets, the earthworms and the sea-worms. The former, including the common soil-makers and a few giants, such as the Tasmanian *Megascolides*, sometimes about six feet long, have bristles but no feet; sense-organs, feelers, and breathing organs are undeveloped as one would expect in subterranean animals. The sea-worms, on the other hand, have usually stump-like bristly feet, and eyes and tentacles and gills, but there is much difference between those which swim freely in the sea (e.g. *Alciope* and *Tomopteris* and some Nereids) and the lobworms which burrow and make countless castings upon the flat sandy shores, or those which inhabit tubes of lime or sandy particles (e.g. *Serpula*, *Spirorbis*, and *Lanice* or *Terebella conchilega*). The earthworms with comparatively few bristles (Oligochæta) are bisexual, while almost all the marine worms with many bristles (Polychæta) have separate sexes. Moreover, those of the first series usually lay their eggs in cocoons, within which the embryos develop without any metamorphosis, while the sea-worms, though they sometimes form cocoons, have free-swimming larvæ usually very different from the adults—little barrel-shaped or pear-shaped ciliated creatures known as Trochospheres.

Some of the Chætopods multiply not only sexually, but asexually by dividing into two or by giving off buds from various parts of their body. Strange branching growths, which eventually separate into individuals, are well illustrated by the freshwater *Nais*, and

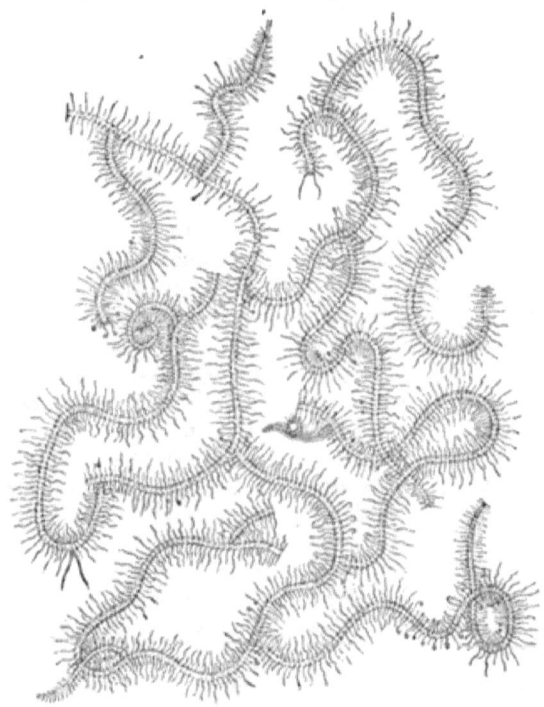

FIG. 43.—A budding marine worm (*Syllis ramosa*). From *Evolution of Sex*; after M'Intosh's *Challenger* Report.)

still better by a marine worm, *Syllis ramosa*, which almost forms a network.

Many sea-worms have much beauty, which some of their names, such as *Nereis, Aphrodite, Alciope*, suggest, and which is said to have induced a specialist to call his seven daughters after them.

Along with the Chætopods, we include some other forms too unfamiliar to find more than mention here, the Myzostomata which form gall-like growths on the feather-stars which they infest, the strange *Bonellia* in which the microscopic male lives as a parasite within the female, and some very simple forms which are sometimes called Archi-Annelids.

8th Class, Hirudinea or Discophora or Leeches.—These are blood-sucking animals, which often cling for a long time to their victims. They live in salt and in fresh water, and sometimes on land. The body is elastic and ringed, but the external markings do not correspond to the internal segments. There are no legs, but the mouth is suctorial, and there is another adhesive sucker posteriorly. The body-cavity is almost obliterated by a growth of spongy tissue, whereas that of Chætopods is roomy. Leeches are hermaphrodite, and lay their eggs in cocoons, within which the young develop without metamorphosis.

The medicinal leeches (*Hirudo medicinalis*) live in slow streams and marshes, creeping about with their suckers or sometimes swimming lithely, preying upon fishes and amphibians, and both larger and smaller animals. They fix themselves firmly, bite with their three semicircular saw-like tooth-plates, and gorge themselves with blood. When they get an opportunity they make the most of it, filling the many pockets of their food-canal. The blood is kept from coagulating by means of a secretion, and on its store the leech may live for many months.

The horse-leech (*Hæmopis sanguisuga*) is common in Britain and elsewhere. The voracious *Aulastoma* is rather carnivorous than parasitic. The land-leeches (e.g. *Hæmadipsa ceylonica*), though small and thin, are very troublesome, sucking the blood of man and beast. Among the others are the eight-eyed *Nephelis* of our ponds, the little *Clepsine* which sometimes is found with its young attached to it, the warty marine *Pontobdella* which fastens on rays, *Piscicola* on perch and carp, *Branchellion* with numerous lateral leaflets of skin, and the largest leech—the South American *Macrobdella valdiviana* which is said to attain a length of over two feet.

Possibly related to the Annelid series are two other classes—

- 9th Class—Chætognatha, including two genera of small arrow-like marine "worms," *Sagitta* and *Spadella*.
- 10th Class—Rotifera, "wheel animalcules," abundant and exquisitely beautiful animals inhabiting fresh and salt water and damp moss. The head-region bears a ciliated structure, whose activity produces the impression of a swiftly rotating wheel. Many of them seem to be entirely parthenogenetic. Some can survive being made as dry as dust.

Fifth set of Worms—a doubtful combination including—

- 11th Class—Sipunculoidea, "spoon-worms" living in the sea, freely or in tubes, e.g. *Sipunculus*.
- 12th Class—Phoronidea, including one genus, *Phoronis*.

13th Class—Polyzoa or Bryozoa, with one exception forming colonies by budding, in fresh water or in the sea, *e.g.* the common sea-mats or horn-wracks (*Flustra*).

14th Class—Brachiopoda or Lamp-shells, a class of marine shelled animals once much richer in members, now decadent. They have a superficial, but only a superficial, resemblance to Molluscs.

I have not catalogued all these classes of "worms" without a purpose. To ignore their diversity would have lent a false simplicity to our survey. If you gain only this idea that there is a great variety—a mob—of worm-like animals, which zoologists have not yet reduced to order, you have gained a true idea. The "worms" lie as it were in a central pool among backboneless animals, from which have flowed many streams of progressive life. They have affinities with Echinoderms, with Insects, with Molluscs, with Vertebrates.

To practical people the study of "worms" has no little interest. The work of earthworms is pre-eminently important; the sea-worms are often used as bait; the leech was once the physician's constant companion; numerous parasitic worms injure man, his domesticated stock, and the crops of his fields.

4. **Echinodermata.**—In contrast to the "Worms," the series including starfishes, brittle-stars, feather-stars, sea-urchins, and sea-cucumbers, is well defined.

The Echinodermata are often ranked next the stinging animals, mainly because many of the adults have a radiate symmetry, as jellyfishes and sea-anemones have. But radiate symmetry is a superficial character, perhaps originally due to a sedentary habit of life in which all sides of the animal were equally affected. Moreover, the larvæ of Echinoderms are bilaterally symmetrical, that is to say, they are divisible into halves along a median plane. We place Echinoderms after and not before "worms," because the simplest worm-like animals are much simpler, much nearer the hypothetical gastrula-like ancestor than are any Echinoderms, and also because it is likely that Echinoderms originated from some worm type or other.

Haeckel used to hold a theory of the starfish which was in some ways suggestive. You know the five-rayed appearance of the animal like a conventional star; you have perhaps watched it moving slowly in a deep rock pool by the shore; you have perhaps discovered that it will surrender one of its arms when you try to capture it. Now Haeckel compared the starfish to a colony of five worms united in the centre. Each "arm" or "ray" is complete in itself. Each has a nerve-cord along the ventral surface, a little eye at the tip, prolongations of the food-canal, blood-vessels, and reproductive organs. Each is anatomically comparable to a worm. Furthermore, when an arm is separated, it may bud out other four arms and thus recreate an entire starfish. Each arm has therefore some physiological independence.

But there is no likelihood that a starfish arose as a colony of worms; the facts of development do not corroborate the suggestion.

Of Echinoderms there are seven classes, two of which are wholly extinct. These—the Cystoids and Blastoids—are of great interest because of their relationship with the feather-stars or Crinoids, which stand somewhat apart from the other four extant classes. The Cystoids are more primitive than the Crinoids, and connect them with the starfishes or Asteroids. The Asteroids are nearly related to the brittle-stars or Ophiuroids, and they are also linked to the sea-urchins or Echinoids. These in turn are the nearest allies of the Holothuroids or sea-cucumbers.

The Echinoderms are all marine. The sea-urchins and Holothurians are mud-cleansing scavengers; the Holothurians and Crinoids feed for the most part on small organisms, though the former are sometimes mud-eaters; the starfishes are more emphatically carnivorous, and often engulf small molluscs.

Among starfishes, sea-urchins, and sea-cucumbers, we find occasional cases of prolonged external connection between the mothers and the young.

The Echinoderms are sluggish animals, though many brittle-stars are lithe gymnasts, and though the commonest Crinoids (Comatulids, such as the rosy feather-star, *Antedon rosacea*), differ from their stalked relatives and adolescent stages in being to some extent swimmers. Perhaps the sluggishness is expressed in the abundance of lime in the skin and other parts; for, as the name suggests, the Echinoderms are thorny-skinned, being usually protected by calcareous plates and spines. The sea-cucumbers are the most muscular and the least limy, indeed in some almost the only calcareous parts are a few anchors and plates scattered in the skin.

Another frequent characteristic is the radial symmetry, but we remember that the larvæ are bilateral.

Very important is the development of a peculiar system of canals and suctorial "tube-feet"—the water-vascular system. By means of the tube-feet the starfishes and sea-urchins move, in the others their chief use seems to be in connection with respiration, and it is likely that in some at least they also help in excretion.

Another characteristic of the Echinoderms is the strangeness of the larval forms. For not only are they very different from the parents, and very remarkable in form, but in no case do they grow directly into the adult. The development is "indirect," the larva does not become the adult; the foundations of the

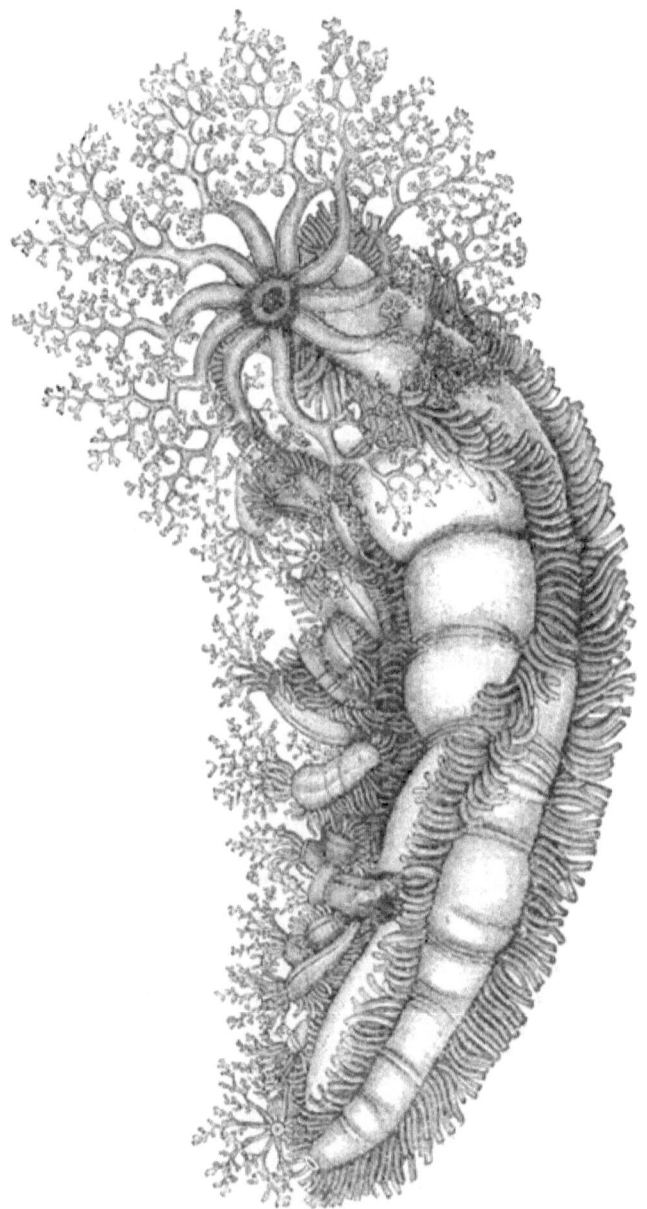

Fig. 44.—A Holothurian (*Cucumaria crocea*) with its young attached to its skin. (From *Evolution of Sex*; after *Challenger* Narrative.)

adult are laid anew within the body of the larva, which is absorbed or partly rejected.

Not only the starfishes but also the brittle-stars and the feather-stars often surrender their arms when captured, or even when slightly irritated, and a part or a remnant can in favourable conditions regrow the whole. The Holothurian *Synapta* breaks readily into pieces, and others contract themselves so forcibly that the internal organs are extruded.

The relations of Echinoderms to other animals are many. A little fish, *Fierasfer*, goes in and out of Holothurians; the degenerate Myzostomata form galls on the arms of Crinoids; starfishes are deadly enemies of oysters. On the other hand, some sea-snails and fishes prey upon Echinoderms in spite of their grittiness. Except that the unlaid eggs of some sea-urchins are edible, and that some sea-cucumbers are considered delicacies, the Echinoderms hardly come into direct contact with human life.

5. **Arthropods.**—Lobsters, centipedes, insects, spiders, agree with the Annelid "worms" in being built up of a series of rings or segments. Some or all of these segments bear limbs, and these limbs are jointed, as the term Arthropod implies. The skin forms an external sheath or cuticle of a stuff called chitin, and this firm sheath helps us to understand how the limbs became well-jointed. The chitin seems in some way antagonistic to the occurrence of ciliated cells, for none seem to occur in this large series unless it be in the strange type *Peripatus*. The chitin has also to do with the moulting or cuticle-casting which is common in the series, for the cuticle is generally rigid and does not expand as the body grows, hence it has to be cast and a new one made. Finally, Arthropods have a nervous system like that of Annelids— a double dorsal brain connected by a ring round the gullet with a double chain of ganglia along the ventral surface. But the life of most Arthropods is more highly pitched than that of Annelids. The sense-organs are more highly developed, brains are larger and more complex, the ganglia of the ventral chain tend to become concentrated; there is division of labour among the appendages; there are new internal organs such as a heart; the whole body is better knit together. A crayfish may part with his claw and grow another in its place, but the animal will not survive being cut in two as some kinds of Annelids do.

The series includes at least five classes:—

    Crustacea, almost all aquatic, and breathing by gills.
    Protracheata, represented by the genus *Peripatus*.
    Myriapoda, centipedes and millipedes.
    Insecta, more or less aerial.
    Arachnida, spiders, scorpions, mites, etc.

The members of the last four classes usually breathe by means of air-tubes or tracheæ, which penetrate into every part of the body, or in the case of spiders and scorpions, by "lung-books," which seem like concentrated and plaited tracheæ. The King-crab (*Limulus*), which is very often ranked along with Arachnids, is aquatic, and breathes by peculiar "gill-books."

(*a*) **Crustacea.**—Except the wood-lice, which live under bark and stones, the land-crabs which visit the sea only at the breeding

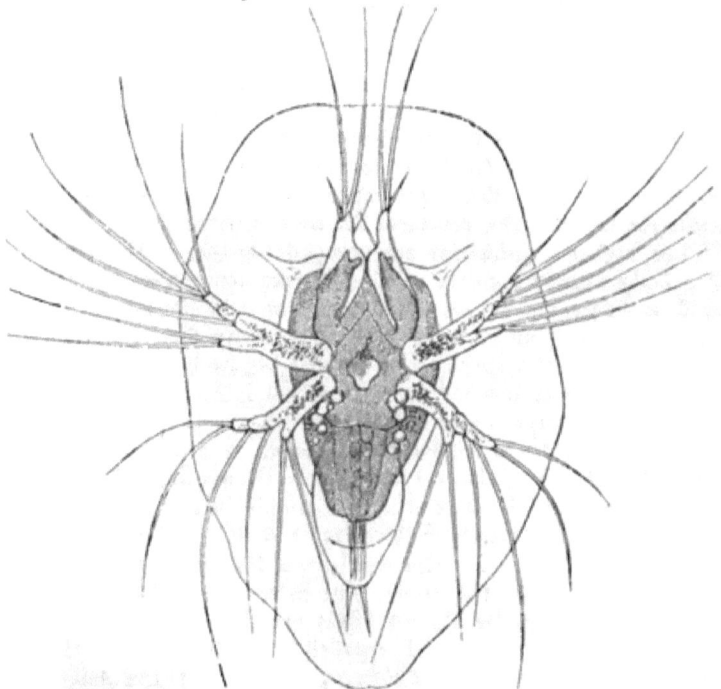

FIG. 45.—Nauplius of Sacculina. (From Fritz Müller.)

time, and some shore-forms which live in great part above the tide-mark, the Crustaceans are aquatic animals, and usually breathe by gills. Each segment of the body usually bears a pair of append-ages, and each appendage is typically double. Among these ap-pendages much division of labour is often exhibited, some being sensory, others masticatory, others locomotor. In the higher forms the life-history is often long and circuitous, with a succession of larval stages.

The lower Crustaceans are grouped together as Entomostraca. They are often small and simple in structure; the number of segments and appendages varies greatly. The little larva which hatches from the egg is usually a "Nauplius"—an unsegmented creature with only three pairs of appendages and a median eye.

The brine-shrimps (*Artemia*), the related genus *Branchipus*, the old-fashioned freshwater *Apus*; the common water-flea *Daphnia* and its relatives, like *Leptodora* and *Moina*, are united in the order of Phyllopods.

The small "water-fleas" of which *Cypris* is a very common representative, and which are very abundant in sea and lake, form the order of Ostracods.

Another "water-flea" *Cyclops* and many more or less degenerate "fish-lice" and other ectoparasites (e.g. *Chondracanthus*, *Caligus*, *Lernæa*) are known as Copepods. The free-swimming forms often occur in great swarms and are devoured by fishes.

The acorn-shells (*Balanus*) crusting the rocks, the barnacles (*Lepas*) pendent from floating "timber," and the degenerate *Sacculina* under the tail of crabs, represent the order Cirripedia.

The higher Crustaceans are grouped together as Malacostraca. The body usually consists of nineteen segments, five forming the head, eight the thorax, six the abdomen or tail. In most cases the larva is hatched at a higher level of structure than the Nauplius represents, but the shrimp-like *Penæus* begins life as a Nauplius while the crab is hatched as a Zoea, the lobster in a yet higher form, and the crayfish as a miniature adult.

Simplest of these higher Crustaceans, in some ways like a survivor of their hypothetical ancestors, is the marine genus *Nebalia*, but we are more familiar with the Amphipods (e.g. *Gammarus*) which jerk themselves along sideways or shelter under stones both in fresh and salt water. The wood-louse *Oniscus* has counterparts (*Asellus*, *Idotea*) on the shore, and several remarkable parasitic relatives. Among the highest forms are the long-tailed lobsters (*Homarus*, *Palinurus*), and crayfishes (*Astacus*), and shrimps (*Crangon*), and prawns (*Palæmon*, *Pandalus*); the soft-tailed hermit crabs (*Pagurus*); and the short-tailed crabs (e.g. *Cancer*, *Carcinus*, *Dromia*).

(*b*) **Protracheata.**—*Peripatus*. This remarkable genus, repre-

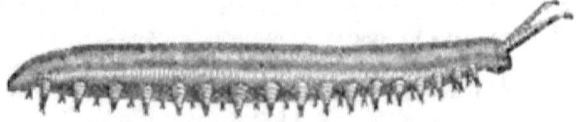

FIG. 46.—Peripatus. (From Chambers's *Encyclop.*; after Moseley.)

sented by about a dozen widely-distributed species, seems to be a survivor of the ancestral insects. Worm-like or caterpillar-like in appearance, with a soft and beautiful skin, with unjointed legs, with the halves of the ventral nerve-cord far apart, and with many other remarkable features, it has for us this special interest that it possesses the air-tubes characteristic of insects and also little kidney-tubes similar to those of Annelids.

(*c*) **Myriapoda.**—Centipedes and Millipedes.—These animals have very uniform bodies, there is little division of labour among the numerous appendages. The head is distinct, and bears besides the pair of antennæ (which *Peripatus* and Insects also have) two pairs of jaws. The Centipedes are flattened, carnivorous, and poisonous; the Millipedes are cylindrical, vegetarian, and innocuous; moreover, they have two pairs of legs to most of their segments.

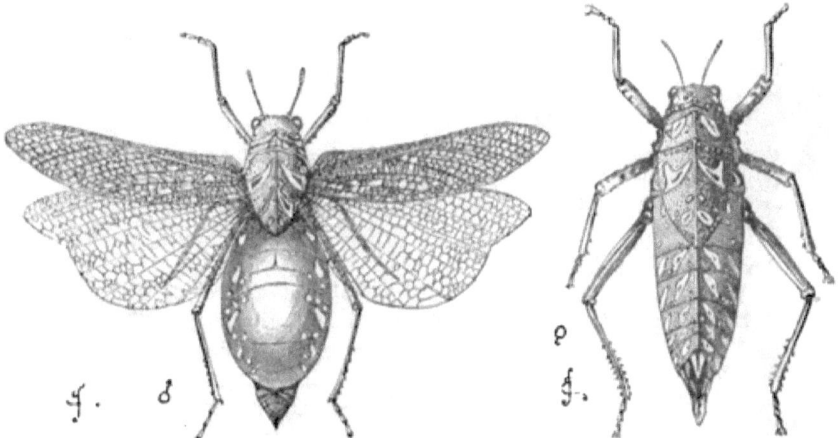

Fig. 47.—Winged male and wingless female of Pneumora, a kind of grasshopper. (From Darwin.)

(*d*) **Insecta.**—Insects are the birds of the backboneless series. Like birds they are on an average active, most have the power of flight, many are gaily coloured, sense-organs and brains are often highly developed.

Contrasted with *Peripatus* and Myriapods, they have a more compact body, with fewer but more efficient limbs. They are Arthropods, which are usually winged in adult life, breathe air by means of tracheæ, and have frequently a metamorphosis in their life-history. To this definition must be added the anatomical facts that the adult body is divided into three regions, (1) a head with three pairs of mouth-appendages (=legs) and a pair of sensitive outgrowths (antennæ or feelers) in front of the mouth, (2) a thorax with three pairs of walking legs, and usually two pairs of wings, and (3) an abdomen without appendages, unless occasional stings, egg-laying organs, etc., be remnants of these.

¹ The wings are very characteristic. They are flattened sacs of skin, into which air-tubes, blood-spaces, and nerves extend. It is possible that they had originally a respiratory, rather than a locomotor function, and that increased activity induced by bettered respiration made them into flying wings.

The breathing is effected by means of the numerous air-tubes or tracheæ which open externally on the sides and send branches to every corner of the body. As the air is thus taken to all the tissues, the blood-vascular system has little definiteness, though there is (as in other Arthropods) a dorsal contractile heart. The larvæ of some insects, *e.g.* dragonflies, mayflies, etc., live in the water, and the tracheæ cannot open to the exterior (else the creature would drown), but they are sometimes spread out on wing-like flaps of skin ("tracheal gills"), or arranged around the terminal portion of the food-canal in which currents of water are kept up.

The student should learn something about the different mouth-organs of insects and the kinds of food which they eat; about the various modes of locomotion, for insects "walk, run, and jump with the quadrupeds, fly with the birds, glide with the serpents, and swim with the fish;" about the bright colours of many, and the development of their senses.

In the simplest insects—the old-fashioned wingless Thysanura and Collembola—the young creature which escapes from the egg-shell is a miniature adult. There is no metamorphosis. So with cockroaches and locusts, lice and bugs; except that the young are small, have undeveloped reproductive organs, and have no wings, they are like the parents, and all the more when the parents (e.g. lice) also are wingless.

In cicadas there is a slight but instructive difference between larvæ and adults. The full-grown insects live among herbage, the young live in the ground, and the anterior legs of the larvæ are adapted for burrowing. Moreover, the larval life ends in a sleep from which an adult awakes. But much more marked is the difference between the aquatic larvæ of mayflies and dragonflies and the aerial adults, in which we have an instance of more thorough though still incomplete metamorphosis.

Different, however, is the life of all higher insects—butterflies and beetles, flies and bees. From the egg-shell there emerges a larva (maggot, grub, or caterpillar), which often lives an active voracious life, growing much, and moulting often. Rich in stores of fatty food, it falls into a longer quiescence than that associated with previous moults and becomes a pupa, nymph, or chrysalis. In this stage, often within the shelter of a silken cocoon, great transformations occur; the body is undone and rebuilt, wings bud out, the appendages of the adult are formed, and out of the pupal husk there emerges an imago, an insect fully formed.

(*e*) **Arachnida.**—Spiders, Scorpions, Mites, etc.—This class is unsatisfactorily large and heterogeneous. In many the body is divided into two regions, the head and breast (cephalothorax), with two pairs of mouth parts and four pairs of walking legs, the abdomen with no appendages. Respiration may be effected by the skin in some mites, by tracheæ in other mites, by tracheæ plus "lung-books" in many spiders, by "lung-books" alone in other spiders, by "gill-books" in the divergent king-crab.

The scorpions with a poisoning weapon at the tip of the tail, the little book-scorpions (*Chelifer*), the long-legged harvest-men (e.g. *Phalangium*); the spiders proper—spinners, nest-makers, hunters; the mites; the strange parasite (*Pentastomum*) in the dog's nose; the quaint king-crab (*Limulus*)—last of a lost race, with which the ancient Trilobites and Eurypterids were connected; all these are usually ranked as Arachnids!

6. **Molluscs.**—It seems strange that animals, the majority of which are provided with hard shells of lime, should be called *mollusca*; for that term first used by Linnæus is a Latinised version of the Greek *malakia*, which means soft. Aristotle applied it originally to the cuttlefish, which are practically without shells, so that its first use was natural enough, but the subsequent history of the word has been strange.

Cockle, mussel, clam, and oyster; snail and slug, whelk and limpet; octopus, squid, and pearly nautilus; what common characteristics have they? Most of them have a bias towards sluggishness, and on the shields of lime which most of them bear, do we not read the legend, "castles of indolence"? But this sluggishness is only an average character, and the shell often thins away. The scallop (*Pecten*) and the swimming *Lima* are active compared with the oyster, and they have thinner shells; the snails which creep slowly between tides or on the floor of the sea are heavily weighted, while the sea-butterflies (Pteropods) have light shells, and most cuttlefish have none at all.

The shell is very distinctive, but we are not able to state definitely how it is formed or what it means. In most of the embryo molluscs which have been studied there is a little pit or "shell-gland" in which a shell begins to be formed, but the shell of the adult is in all cases made by a single or double fold of skin known as the "mantle." In some cases where the shell seems to be absent, *e.g.* in some slugs, a degenerate remnant is still to be found beneath the skin, while in other cases (*e.g.* most cuttlefish) its absence is to be explained as a loss, since related ancestral species possess it. There are, however, two or three primitive forms

(*Chætoderma, Neomenia*) where an incipient shell is represented only by a few spicules or plates of lime.

The shell is made by the folded skin or "mantle"; it consists for the most part of carbonate of lime along with a complex organic substance called conchiolin; it shows three layers, of which the outermost is somewhat soft and without lime, while the innermost shines with mother-of-pearl iridescence. The whole product is a cuticle—something formed from the skin; its varied colours and forms are beautiful; it is a protective shield; but there are many

FIG. 48.—The common octopus. (From Chambers's *Encyclop.*; after Brehm.)

questions about shells which we cannot answer. Where does the carbonate of lime come from, since that salt is often far from abundant in the water in which most molluscs live? Have they the power of changing the abundant sulphate of lime in sea-water into carbonate of lime, perhaps by an interaction with waste products excreted from the skin? Is the shell an expression of the constitutional sluggishness of the animal since it seems on the whole to be most massive in the most sluggish, least so in the most active forms?

Most molluscs are marine, on the shore, in the open sea, in the great depths; there are also many freshwater forms, *e.g.* the mussels *Anodon* and *Unio*, and the snails, *Lymnæus*, *Planorbis*, and *Paludina*; the terrestrial snails and slugs are legion. Among those of the shore the naked Nudibranchs are often in colour and form protectively adapted to their surroundings; those of the open sea (Heteropods, Pteropods, and many cuttlefish) are active and carnivorous, with light shells or none; in the dark depths many are blind or in other ways rudimentary, but food seems to be so abundant that there is almost no need to struggle for it.

As to diet, there are three kinds of eaters—carnivores, such as the active swimmers we have mentioned, besides the whelks and many other burglars who bore through their neighbours' shells, and the *Testacella* slugs; vegetarians, like the periwinkle, the snail, and most slugs; and thirdly, almost all the bivalves, which feed on microscopic plants and animals, and on organic débris wafted to the mouth by the lashing of the cilia on the gills and lips. In this connection it is important to notice that all molluscs except bivalves have in their mouths a rasping ribbon or toothed tongue (*radula, odontophore*) by which they grate, file, or bore with marked effect. Of parasites there are few, but one Gasteropod, *Entoconcha mirabilis*, which lives inside the Holothurian *Synapta*, is very remarkable in its degeneration. It starts in life as a vigorous embryo like that of most marine snails, it becomes a mere sac of reproductive organs and elements.

In structure, molluscs differ remarkably from the arthropods and higher "worms" in the absence of segments and serial appendages. They are not divided into rings, and they have no legs.

To begin with, they were doubtless (bilaterally) symmetrical animals, and this symmetry is retained in primitive forms like the eight-shelled *Chiton* and in the bivalves. But most of the snails are twisted and lop-sided, they cannot be symmetrically halved. For this asymmetry the strange dorsal hump formed by the viscera, and the tendency that the single shell would have to fall to one side, are sometimes blamed. That this lop-sidedness is not necessarily a defect, but rather the reverse, is evident from the success not only of the snail tribe but of many other asymmetrical animals.

The skin has a remarkable fold (double in the bivalves) known as the "mantle," the importance of which in making the shell we have already recognised. Another very characteristic structure is the so-called "foot," a muscular protrusion of the ventral surface, an organ used in creeping and swimming, leaping and boring, but almost absent in the sedentary oysters.

We rank the molluscs high among backboneless animals, partly because of the nervous system, which here as elsewhere is a dominating characteristic. There are fewer nerve centres than in most Arthropods or in higher "worms," but this is in most cases a sign of concentration. There is a (*cerebral*) pair with nerves which supply the head region, another (*pleural*) pair with nerves to the sides and **viscera**, a third (*pedal*) pair whose nerves govern the foot, and often other accessory centres of which the most important are *visceral*. In the somewhat primitive eight-shelled *Chiton* and its neighbours, the nervous system is the most readily harmonised with that of other Invertebrates ; in **bivalves** the three pairs of centres are far apart ; in most snails and in **cuttlefish** the three are concentrated in the head-regions, and it is those forms with concentrated ganglia which show some signs of cleverness and emotion.

Life-History.—Most molluscs pass through two **larval stages** before they acquire their characteristic adult appearance. The first is interesting because it is virtually the same as the young stage of many "worms." It is a barrel-shaped or pear-like embryo with a ring of locomotor cilia in front of the mouth, and is known as a *Trochosphere*.

After a while this changes into a more characteristic form called the *Veliger*. It bears on its head a ciliated cushion or velum often produced into lobes ; the body has a ventral "foot" and a dorsal "shell-gland." In aquatic Gasteropods the visceral hump begins to appear at this stage.

The eggs of cuttlefish differ from those of other molluscs in their rich supply of yolk, which serves for a prolonged period as capital for the young, and the two larval stages noticed above are skipped over. There are other interesting modifications in the life-history of terrestrial and freshwater forms, witness the little larvæ of the freshwater mussel which are kept within the gills of the mother till the approach of some sticklebacks or other fish, to which the liberated young then fix themselves.

History.—The shells of molluscs are well preserved in the rocks, and palæontologists have been very successful in tracing long series.

The chief types are all represented in the Cambrian strata—a fact which forcibly suggests the immensity of yet earlier ages. They are abundant from the Silurian onwards. The snails have gone on increasing, and are now more abundant than ever ; the bivalves cannot be said to have diminished, but the Cephalopod tribe has dwindled. Of the Nautiloid type of Cephalopods, of which there are crowds of fossil forms, only the pearly Nautilus now survives, and though there are many kinds of naked cuttle-fish in our seas there are ten times as many shelled ancestors in the rocks. The geological record confirms what we should otherwise expect, that the lung-breathing snails and the freshwater bivalves were somewhat late in appearing.

Prof. Ray Lankester has reconstructed an ideal ancestor which combines the various molluscan characteristics in a satisfactory fashion, and may be something like the original mollusc. Whence that original sprang is uncertain, but the common occurrence of the trochosphere larva and some of the characters of the primitive Gasteropods (*Neomenia, Chætoderma, Chiton*) suggest the origin of molluscs from some "worm" type or other. We can be sure of this, however, that the series must have divided at a very early epoch into two sets, the sluggish, sedentary, headless bivalves on the one hand, and the more active and aggressive snails and cuttle-fish on the other.

Relation to Man.—Irresistibly we think first of oysters, which Huxley describes as "gustatory flashes of summer lightning," and over which neolithic man smacked his lips. But many others, cuttlefish, ear-shells (*Haliotis*), mussels (*Mytilus edulis*), peri-winkles (*Littorina littorea*), cockles (*Cardium cordatum*), etc. etc., are used as food, and many more as bait. In ancient days, as even now, the shells of many were used for ornaments, instruments, lamps, vessels, coins, etc.; the inner layer of the shell furnishes mother-of-pearl; concretions around irritating particles become pearls in the pearl-oyster (*Margaritana*) and in a few others; the Tyrian purple was a secretion of the whelk *Purpura* and the related *Murex*; and the attaching byssus threads of the large bivalve *Pinna* may be woven like silk.

On the other hand, a few cuttlefish are large enough to be somewhat dangerous; the bivalve *Teredo* boring into ship-bottoms and piers is a formidable pest, baulked, however, by the pre-valent use of metal sheathing; the snails and slugs are even more voracious than the birds which decimate them.

Conchology was for a while a craze, rare shells have changed hands at the cost of hundreds of pounds, such is the human "mania of owning things." But the shells are often fascinating in their beauty, and poetic fancy has played lovingly with such as the Nautilus.

CHAPTER XVI

BACKBONED ANIMALS

1. *Balanoglossus*—2. *Tunicates*—3. *The Lancelet*—4. *Round-Mouths or Cyclostomata*—5. *Fishes*—6. *Amphibians*—7. *Reptiles*—8. *Birds*—9. *Mammals*

ACCORDING to Aristotle, fishes and all higher animals were "blood-containing," and thus distinguished from the lower animals, which he regarded as "bloodless." He was mistaken as to the absence of blood in lower animals, for in most it is present, but the line which he drew between higher and lower animals has been recognised in all subsequent classifications. Fishes, amphibians, reptiles, birds, and mammals differ markedly from molluscs, insects, crustaceans, "worms," and yet simpler animals. The former are backboned (Vertebrate), the latter backboneless (Invertebrate).

It is necessary to make the contrast more precise. (*a*) Many Invertebrates have a well-developed nerve-cord, but this lies on the ventral surface of the body, and is connected anteriorly, by a ring round the gullet, with a dorsal brain in the head. In Vertebrates the whole of the central nervous system lies along the dorsal

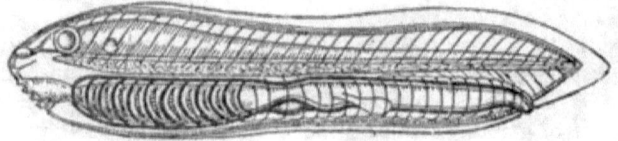

FIG. 49.—Diagram of "Ideal Vertebrate," showing the segments of the body, the spinal cord, the notochord, the gill-clefts, the ventral heart. (After Haeckel.)

surface of the body, forming the brain and spinal cord. These arise by the infolding of a skin groove on the dorsal surface of the embryo. (*b*) Underneath the nerve-cord in the Vertebrate embryo is a supporting rod or notochord. It arises along the roof of the food-canal, and serves as a supporting axis to the body. It persists in some of the lowest Vertebrates (*e.g.* the lancelet); it persists in part in some fishes; but in most Vertebrates it is replaced by a new growth—the backbone—which ensheaths and constricts it. (*c*) From the anterior region of the food-canal in fishes and tadpoles slits, bordered by gills, open to the exterior. Through the slits water flows, washing the outsides of blood-vessels and aërating the blood. These slits or clefts are represented in the young of all Vertebrate animals, but in reptiles, birds, and mammals they are transitory and never used. Amphibians are the highest animals in which they are used for breathing, and even then they may be entirely replaced by lungs in adult life. They are evident in tadpoles, they have disappeared in frogs. (*d*) Many an Invertebrate has a well-developed heart, but this always lies on the dorsal surface of the body, while that of fish or frog, bird or man, lies ventrally. (*e*) It is characteristic of the eye of backboned animals that the greater part of it arises as an outgrowth from the brain, while that of backboneless animals is directly derived from the skin. But this difference is less striking when we remember that it is from an infolding of skin that the brain of a backboned animal arises.

But while the characteristics of backboned animals can now be stated with a precision greater than that of sixty years ago, it is no longer possible to draw with a firm hand the dividing line between backboned and backboneless. Thus fishes are not the simplest Vertebrates; the lamprey and the glutinous hag belong to a more primitive type, and are called fishes only by courtesy; simpler still is the lancelet; the Tunicates hesitate on the border line, being tadpole-like in their youth, but mostly degenerate when adults; and the worm-like *Balanoglossus* is perhaps to be ranked as an incipient Vertebrate. The extension of knowledge and the application of evolutionary conceptions obliterate the ancient landmarks of more rigid but less natural classification.

1. **Balanoglossus.**—*Balanoglossus* is a worm-like animal, represented by some half-dozen species, which eat their way

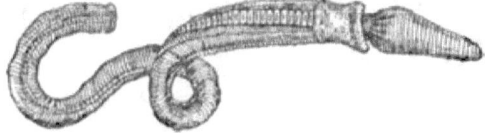

FIG. 50.—Balanoglossus, showing proboscis, collar, and gill-slits.

through sandy mud off the coasts of the Channel Islands, Brittany, Chesapeake Bay, and other regions. Its body is ciliated and divided into distinct regions—a large "proboscis" in front of the mouth, a firm collar behind the mouth, a part with numerous gill-slits behind the collar, and finally a soft coiled portion with the intestine and reproductive organs. The size varies from about an inch to 6 inches, the colours are bright, the odour is peculiar; the sexes are separate. But *Balanoglossus* is most remarkable in having a dorsal supporting rod (like a notochord) in the "proboscis" region, a dorsal nerve-cord running along the back and especially developed in the collar, and a series of gill-clefts on the anterior part of the food-canal. It is therefore difficult to exclude *Balanoglossus* from

FIG. 51.—Cephalodiscus, a single individual, isolated from a colony. It is much magnified. (From Chambers's *Encyclop.*; after *Challenger* Report, by M'Intosh and Harmer.)

the Vertebrate series, and it is likely that the same must be said of another strange animal, *Cephalodiscus*, discovered by the *Challenger* explorers.

2. **Tunicates.**—Hanging to the pennon-like seaweeds which fringe the rocky shore and are rarely uncovered by the tide, large sea-squirts sometimes live. They are shaped like double-mouthed wine-bags, 2 or 3 inches in length, and water is always being drawn in at one aperture and expelled at the other. Usually they live in clusters, and their life is very passive. We call them sea-squirts because water may spout forth when we squeeze their

bodies, while the title Tunicate refers to a characteristic cloak or tunic which envelops the whole animal.

There is not much to suggest backbonedness about these Tunicates, and till 1866 no one dreamt that they could be included in the Vertebrate series. But then the Russian naturalist Kowalevsky discovered their life-history. The young forms are free-swimming creatures like miniature tadpoles, with a dorsal nerve-cord, a supporting rod in the tail region, gill-slits opening from the food canal, a little eye arising as an outgrowth of the brain, and a ventral heart.

There are only two or three genera of Tunicates, especially one called *Appendicularia*, in which these Vertebrate characteristics are retained throughout life. The others lose them more or less completely. The young Tunicates are active, perhaps too active, for a short time ; then they settle down as if fatigued, fix themselves by their heads, absorb their tails, and become deformed. The nervous system is reduced to a single ganglion between the two apertures ; the original gill-slits are replaced by a great number of a different character ; the eye is lost. From the skin of the degenerate animal the external tunic is exuded. It is a cuticle, and consists, in part at least, of cellulose, the substance which forms the cell-walls of plants. Thus this characteristically vegetable substance occurs almost uniquely in the most passive part of a very passive animal. The sea-squirt's metamorphosis is one of the most signal instances of degeneration ; the larva has a higher structure than the adult ; the young Tunicate is a Vertebrate, the adult is a nondescript. We cannot tell how this fate has befallen the majority, nor why a few are free-swimmers, nor why *Appendicularia* retains throughout life the Vertebrate characteristics of its youth. Do the majority overexert themselves when they are "tadpoles," or are they constitutionally doomed to become sedentary?

Tunicates are hermaphrodite—a very rare condition among Vertebrates ; some of them exhibit "alternation of generations," as the poet Chamisso first observed ; asexual multiplication by budding is very common, and not only clusters but more or less intimate colonies are thus formed.

Tunicates live in all seas, mostly near the coast from low water to 20 fathoms, and usually fixed to stones and rocks, shells and seaweed. A few are free-swimming, such as the fire-flame (*Pyrosoma*), a unified colony of tubular form, sometimes 2 or 3 feet in length, and brilliantly phosphorescent. Very beautiful are the swimming chains of the genus *Salpa*, whose structure and life-history alike are complicated.

Tunicates feed on the animalcules borne in by the water currents, and some of them must feed well, so rapidly do they grow and multiply. Unpleasant to taste, they are left in peace, though a crab sometimes cuts a tunic off as a cloak for himself.

3. **The Lancelet.**—The lancelet (*Amphioxus*) is a simple Vertebrate, far below the **structural rank** of fishes. It is only about 2 inches in length, and, **as both English** and Greek names suggest, it is pointed at both ends. On the **sandy** coasts of warm and temperate **seas** it is widely distributed.

From tip to tail of the translucent body runs a supporting notochord; above this is a spinal cord, with hardly a hint of brain. The pharynx bears a hundred or so gill-slits, which in the adult are covered over by folds of skin, so that the water which enters by the mouth finds its way out by a single posterior aperture. Although *Amphioxus* has no skull, nor jaws, nor brain, nor limbs, it deserves its position near the base of the Vertebrate series. The sexes are separate, and the eggs are fertilised outside of **the** body. The development of the embryo has been very carefully studied, and is for a time very like that of Tunicates.

4. **Round-Mouths or Cyclostomata.**—The hag-fishes and the lampreys and a few allied genera must be excluded from the class of fishes. They **are** survivors of a more primitive race. They are jawless, limbless, scaleless, and therefore not fishes.

The lampreys (*Petromyzon*) live in rivers and estuaries, and also in **the** wider sea. They are eel-like, slimy animals. The skeleton is gristly; the simple brain is imperfectly roofed; the single nostril does not open into the mouth; the rounded mouth has horny teeth on the lips and on the piston-like tongue; there **are seven** pairs of gill-pouches which open directly to the exterior and internally into a tube lying beneath and communicating with the adult gullet; the young are blind and otherwise different from the parents, and may remain so for **two** or three years.

Though lampreys eat worms and other small fry, and even dead animals, they fix themselves aggressively **to** fishes, rasping holes in **the** skin, and sucking the flesh **and** juices. They also cling to stones, as the name *Petromyzon* suggests.

Some species drag stones into **a** kind of nest. They spawn in spring, usually far up rivers, for at least some of the marine lampreys leave the sea at the time of breeding. The young are in many ways different from the parents, and that of the small river lampern (*Petromyzon branchialis*) used to be regarded as a distinct animal—*Ammocœtes*. The metamorphosis was discovered two hundred years ago by Baldner, a Strasburg fisherman, but was overlooked till the strange story was worked out in 1856 by August Müller. Country boys often call the young "nine-eyes," miscounting the gill apertures, and **the** Germans also speak of *neun-augen*.

The sea lamprey (*P. marinus*) may measure three feet; the river lamprey (*P. fluviatilis*) about two feet; the small lampern or stone-grig (*P. branchialis* or *planeri*) about a foot. The flesh is well known to be palatable.

The glutinous hag (*Myxine glutinosa*) is an eel-like animal, about a foot in length, of a livid flesh colour. It is common at considerable depths (40 to 300 fathoms) off the coasts of Britain and Norway, and, when not feeding, lies buried in the mud with only its nostril protruded. Like the lamprey, it has a smooth slimy skin, a gristly skeleton, a round suctorial mouth with teeth. The single nostril communicates with the food-canal at the back of the mouth, and serves for the inflowing of water; the six gill-pockets on each side open directly into the gullet, but each has an excurrent tube, and the six tubes of each side open at a common aperture. The animal lives away from the light, and its eyes are rudimentary, hidden beneath skin and muscles. The skin exudes so much slime that the ancients spoke of the hag "turning water into glue."

In several ways the hag is strange. Thus J. T. Cunningham discovered that it is hermaphrodite, first producing male elements, and afterwards eggs, and Nansen has corroborated this. The eggs are large and oval, each enclosed in a "horny" shell with knotted threads at each end, by which a number are entangled together. How they develop is unknown. The hags devour the bait and even the fish from the fisherman's lines, and some say that they bore their way into living fishes such as cod.

5. **Fishes.**—Fishes are in the water as birds in the air,—swift, buoyant, and graceful. They are the first backboned animals with jaws, while scales, paired fins, and gills are their most characteristic structures. The scales may be hard or soft, scattered or closely fitting, and are often very beautiful in form and colour. The paired fins are limbs, as yet without digits, varying much in size and position, and helping the fish to direct its course. The gills are outgrowths of skin with a plaited surface, on which the branching blood-vessels are washed by the water. They are the breathing organs of all fishes, but in the double-breathing mud-fishes (*Dipnoi*) the swim-bladder has come to serve as a lung, and there are hints of this in a few others.

There are at least four orders of fishes :—

(1) The cartilaginous fishes (Elasmobranchs or Selachians) are for the most part quite gristly, except in teeth and scales. Among them are the flattened skates and rays with enormous fore-fins, while the sharks and dogfish are shaped like most other fishes. Their pedigree goes back as far as the Silurian rocks, in which remains of shark-like forms are found. A Japanese shark (*Chla-*

mydoselachus) is said to be very closely allied to types which occur in the Old Red Sandstone. Allied to the Elasmobranchs, but sometimes kept in a separate division, are two genera, the *Chimæra* or King-of-the-Herrings, and *Callorhynchus*, its relative in Southern Seas.

(2) The Ganoid fishes are almost, if not quite, as ancient as the Elasmobranchs, but their golden age, long since past, was in Devonian and Carboniferous ages. There are only some seven different kinds now alive. Two of these are the sturgeons (*Acipenser*) and the bony pike (*Lepidosteus*). The latter has a bony skeleton; the sturgeon is in part gristly. An armature of hard scales is very characteristic of this decadent order.

(3) In Permian times, when Reptiles were beginning, a third type of fish appeared, of which the Queensland mud-fish (*Ceratodus*) seems to be a direct descendant. In this type the air-bladder is used as a lung, thus suggesting the transition from Fishes to Amphibians. Perhaps this order was always small in numbers; nowadays at least there are only two genera—*Ceratodus*, from the fresh water of Queensland, and *Protopterus*, from west and tropical Africa; while another form, sometimes called a different genus (*Lepidosiren*), is recorded from the Amazons. Double-breathers or Dipnoi we call them, for they do not depend wholly upon gills, but come to the surface and gulp air into their air-bladder. Mud-fishes they are well named, for as the waters dry up they retire into the mud, forming for themselves a sort of nest, within which they lie dormant.

(4) In the Chalk period the characteristically modern fishes (Teleosteans), with completely bony skeletons, began. Herring and salmon, cod and pike, eel and minnow, and most of the commonest fishes belong to this order. Heavy ironclads yield to swift gunboats, and the lithe Teleosteans have succeeded better than the armoured Ganoids.

The wedge-like form of most fishes is well adapted for rapid swimming. Most flat fish, whether flattened from above downwards like the gristly skate, or from side to side like the flounders and plaice, live at the bottom; those of eel-like shape usually wallow in the sand or mud; the quaint globe-fish float passively. The chief organ of locomotion is the tail; the paired fins help to raise or depress the fish, and serve as guiding oars. In the climbing perch they are used in scrambling; in the flying fish they are sometimes moved during the long swooping leaps. In eels and pipe-fish they are absent; in the Dipnoi they have a remarkable median axis. The unpaired fins on the back and tail and under surface are fringes of skin supported by rays.

Fishes are often resplendent in colours, which are partly due to pigments, partly to silvery waste-products in the cells of the outer skin, and partly to the physical structure of the scales. Sometimes the males are much brighter than the females, and grow brilliant at the breeding season. In some cases the colours harmonise with surrounding hues of sand and gravel, coral and seaweed; while the plaice and some others have the power of rapidly changing their tints.

Fishes feed on all sorts of things. Some are carnivorous, others

FIG. 52.—The gemmeous dragonet (*Callionymus lyra*), the male above, the female beneath. (From Darwin.)

vegetarian, others swallow the mud. By most of them worms, crustaceans, insect-larvæ, molluscs, and smaller fishes are greedily eaten. Strange are some of large appetite (*e.g. Chiasmodon niger*), who manage to get outside fishes larger than their own normal size!

Of their mental life little is known. Yet the cunning of trout, the carefulness with which the mother salmon selects a spawning-ground, the way the archer-fish (*Toxotes*) spits upon insects, the nest-making and courtship of the stickleback and others, the pugnacity of many, show that the brain of the fish is by no means asleep.

The **males are** often different from the females—smaller, brighter, and less **numerous.** In some cases they court their mates, and fight with **their** rivals. Most of the females lay eggs, but a few bony fishes and many sharks bring forth living young. In two sharks there is a prophecy of that connection between mother and offspring which is characteristic of **mammals.** The fish's egg is usually a small thing, but those of Elasmobranchs are large, being rich in yolk and often surrounded by a mermaid's purse. This egg-case has long tendril-like prolongations at the corners, these twine automatically around seaweed, **and the** embryos may be rocked by the waves until the time of hatching. When the egg is enclosed in a sheath, **or when** the young are hatched within the body **of** the mother, fertilisation must take place internally, but in most cases the male accompanies the female as she spawns, and with his milt fertilises the eggs in the water or on the gravelly spawning-ground. As love for offspring varies inversely with their number, there is little parental care among the prolific fishes.

Most fishes live either wholly in the sea or wholly in fresh water, but some are indifferent, and pass, at spawning time especially, from one to the other. A few, such as the climbing perch,* venture ashore, while the mud-fishes and a few others can survive drought for a season. In caves several blind fishes live, and species of *Fierasfer* find more or less habitual lodging inside sea-cucumbers and some other animals.

The fishes which live **in** deep water **are** interesting in many ways. Günther has shown that from 80 to 200 fathoms the eyes are rather larger than usual, as if to make the most of the dim light. Beyond 200 fathoms "small-eyed fishes as well as large-eyed occur, the former having their want of vision compensated for by tentacular organs of touch, whilst the latter have no such accessory organs," and can see only by the fitful light of **phosphorescence.** "In the greatest depths blind fishes occur, with rudimentary eyes, and without special organs of touch." The phosphorescence is produced by numerous marine animals and by the fishes themselves.

6. **Amphibians.**—The Amphibians which **now** live are neither numerous nor large. Giant Amphibians or Labyrinthodonts began to appear in the Carboniferous period, but most of the modern frogs and toads, newts and salamanders, are relatively pigmies.

Young Amphibians always breathe by gills, as Fishes do, and in some cases these gills persist in adult life. But whether they do or not, the full-grown Amphibians have lungs and use them. The skin is characteristically soft, naked, and clammy. Amphibians are the first Vertebrates with hands and feet, with fingers and toes. Unpaired fringes are sometimes present on the back and tail as in Fishes, but are never supported by fin-rays.

The class includes four orders, of which the Labyrinthodonts are wholly extinct, the other three being represented by tail-less frogs and toads (Anura), by newts and salamanders (Urodela) with distinct tails, and by a few of worm-like form and burrowing habit, *e.g. Cæcilia*. Some, the last for example, are terrestrial, but usually live in damp places; most pass their youth at least in fresh water; none can endure saltness, and they are therefore absent from almost all oceanic islands. The common British newts (*Triton* and *Lissotriton*), and the often brightly-coloured salamanders (*Salamandra*) have in adult life no trace of gills; the rice-eel (*Amphiuma*) and the genus *Menopoma* lose their gills, but persistent clefts indicate their position; the blanched blind *Proteus* from caves and the genus *Menobranchus* keep their gills throughout life. The remarkable Axolotl from North American lakes occurs in two forms, both of which may bear young; the one form (*Axolotl*) has persistent gills, the other form (*Amblystoma*) loses them, and the change from the *Axolotl* to the *Amblystoma* is in part associated with the passage from the water to the swampy shore. A large fossil discovered by Scheuchzer in the beginning of the eighteenth century was quaintly regarded as a fossil man and as a testimony of the deluge. But Cuvier showed that Scheuchzer's *Homo diluvii testis* was but a large newt.

The common frogs (*Rana*), the Surinam toad (*Pipa*), the common toads (*Bufo*), and the tree-frogs (*Hyla*) illustrate the tailless order Anura. In none of them is there in adult life any trace of gills.

The worm-like, limbless, burrowing Amphibians (Gymnophiona) must not be confused with the blind- or slow-worms, which are lizards. There are only very few genera, *Siphonops*, *Rhinatrema*, *Epicrium*, *Cæcilia*. The newly-born *Cæcilia* has external gills, but these are soon lost. The eyes are covered with skin, but are well developed.

The race of Amphibians began in the Carboniferous ages. Most of the Labyrinthodonts which flourished then and in the two succeeding periods were newt-like in form, but some were serpentine. They seem to have been armoured, and were sometimes large.

Amphibians are naturally sluggish. For long periods they can fast and lie dormant; they can survive being frozen quite stiff, and though tales of toads within stones are mostly due to mistakes or fancies, there are some authentic cases of prolonged imprisonment.

Few are found far from water, and the gilled condition of the young is skipped over only in a few cases. In the black salamander (*Salamandra atra*) of the Alps, which lives where pools are scarce, the young, after living and breathing for a time within the mother, are born as lung-breathers; also in some species of tree-frogs (*Hylodes*), which live in situations where water is scarce, the gilled stage is omitted.

The development of the common frog should be studied by every student of natural history. The eggs are fertilised as they are being laid. The division of the ovum can be readily observed. In its early stages the tadpole is fish-like, with a lamprey-like

FIG. 53. —The life-history of the Frog.

mouth. External gills are replaced by an internal set, and as metamorphosis is accomplished these disappear and the lungs become active. The larva feeds first on its own yolk, then on freshwater plants, then on small animals or even on its own relatives; then it fasts, absorbing its tail, and finally it becomes an insect-catching frog.

The food of adult Amphibians usually consists of insects, slugs, and worms; most of the larvæ are for a time vegetarian. Though Amphibians often live alone, crowds are often found together at the breeding season. Then the sluggish life wakes up, as the croakings of frogs remind us. Quaint are many of their reproductive habits, to some of which allusion has already been made. Such animals as the Surinam toad (*Pipa americana*) and the Obstetric frog (*Alytes obstetricans*) suggest that the Amphibians make experiments in eugenics.

7. **Reptiles.**—Fishes and Amphibians are closely allied; so Reptiles are linked to Birds, and more remotely to Mammals also. Those three highest classes—Reptiles, Birds, and Mammals—are very different from one another, but they have certain characters in common. Most of them have passed from the water to dry land; none of them ever breathe by gills; all of them have two embryonic birth-robes—amnion and allantois—which are of great importance in early life. Compared with the other Vertebrates, the brains are more complex, the circulation is more perfect, the whole life has a higher pitch. As symbols of mammal, bird, and reptile, take the characteristic coverings of the skin—hair, feathers, and scales. Hair typifies strength and perhaps also gentleness; feathers suggest swift flight, the beauty which wins love, and the down which lines the warm nest; scales speak of armour and cold-blooded stealth.

But we need not depreciate reptiles, nor deny the justice of that insight which has found in them the fittest emblems of the omnipotence of the earth. If Athene of the air possesses the birds, surely the power of the dust is in the grovelling snakes. Few colour arrangements are more beautiful than those which adorn the lithe lizards. The tortoise is an example of passive energy, self-contained strength, and all but impenetrable armature. The crocodiles more than the others recall the strong ferocity of the ancient extinct dragons. Nor should we judge reptiles exclusively by their living representatives, any more than we should judge the Romans by those of the decadent Empire. It is interesting to remember the long-tailed toothed *Archæopteryx*, the predecessor of modern birds, just as it is to recall the giant sloths which preceded the modern Edentate mammals; but it is essential to include in our appreciation of Reptiles the giant dragons of their golden age. Most modern forms are pigmies beside an *Ichthyosaurus* 25 feet long, a *Megalosaurus* of 30, a *Titanosaurus* of 60, or an *Atlantosaurus* of 100, all fairly broad in proportion. We have still pythons and crocodiles and other reptiles of huge size, and we do not deny Grant Allen's remark that a good blubbery "right whale" could give points to any deinosaur that ever moved upon Oolitic continents, but the fact remains that in far back times (Triassic, Jurassic, and Cretaceous) reptiles had a golden age with a predominance of forms larger than any living members of the class. Besides size, however, the ancient saurians had another virtue, apparently possessed by both small and great—they were progressive. For, with toothed birds on the one hand and flying or flopping reptiles on the other, it seems probable that birds had their origin from feverish saurians which acquired the power of flight, and it is also possible that some, perhaps pathological, mother reptile, overflowing in the milk of animal kindness, and retaining her young for a long time within her womb, was the forerunner of the mammalian race.

While there are many orders of extinct reptiles—Ichthyosaurs, Plesiosaurs, Deinosaurs, Pterosaurs, and other saurians not yet classified with certainty—the living forms belong to four sets—the lizards, the snakes, the tortoises, and the crocodiles—to which a fifth order should perhaps be added for the New Zealand "lizard" *Hatteria* or *Sphenodon*, which is in several respects a living fossil.

The Lizards (Lacertilia).—The lizards form a central order of Reptiles, but the members are a motley crowd, varied in detailed structure and habit. Usually active in their movements, though fond, too, of lying passive in the sunshine, they are often very beautiful in form and colour, and not uncommonly change their tints in sympathetic response to their surroundings. Most lay eggs, but in some, *e.g.* the common British lizard (*Lacerta* or *Zootoca vivipara*), and the slow-worm, the young are hatched within the mother.

Among the remarkable forms are the Geckos, which with plaited adhesive feet can climb up smooth walls; the large Monitors (*Varanus*), which may attain a length of 6 feet, and prey upon small mammals, birds, frogs, fishes, and eggs; the poisonous Mexican lizard (*Heloderma horridum*), with large venom-glands and somewhat fang-like teeth; the worm-like, limbless *Amphisbæna*; the likewise snake-like slow-worm (*Anguis fragilis*), which well illustrates the tendency lizards have to break in the spasms of capture; the large Iguanas, which frequent tropical American forests, and feed on leaves and fruit; the sluggish and spiny "Horned Toad" (*Phrynosoma*); the Agamas of the Old World comparable to the Iguanas of the New; the Flying Dragon (*Draco volans*), which, with skin outstretched on extended ribs, swoops from tree to tree; the Australian frilled lizards (*Chlamydosaurus*) and the quaint thorny *Moloch*; the single marine lizard (*Oreocephalus* or *Amblyrhynchus cristatus*) from the Galapagos; and the divergent Chamæleons, flushing with changeful colour.

The New Zealand *Hatteria* or *Sphenodon* is quite unique, and seems to be the sole survivor of an extinct order—Rhynchocephalia. It was in it first of all that the pineal body—an upgrowth from the mid-brain of backboned animals—was seen to be a degenerate upward-looking eye.

Snakes or Serpents (Ophidia). — These much modified reptiles mostly cleave to the earth, though there are among them clever climbers, swift swimmers, and powerful burrowers. Though

FIG. 54.—The wall-lizard (*Lacerta muralis*), three differently coloured and marked varieties of the same species. (Compiled from Eimer.)

they are all limbless, unless we credit the little hind claws of some boas and pythons with the title of legs, they flow like swift living streams along the ground, using ribs and scales instead of their lost appendages, pushing themselves forward with jerks so rapid that the movement seems continuous. Without something on which to raise themselves they must remain at least half prostrate, but in the forest or on rough ground there are no lither gymnasts. Their united eyelids give them an unlimited power of staring, and, according to uncritical observers, of fascination; yet most of them seem to see dimly and hear faintly, trusting mainly for guidance to the touch of their restless protrusible tongue and to their sense of smell. Their only language is a hiss or a whine. Most of them have an annual period of torpor, and all periodically cast off their scales in a normally continuous slough, which they turn outside-in as they crawl out. Almost all lay eggs, but in a few cases (*e.g.* the adder) the young are hatched within the mothers, and this mode of birth may be induced by artificial conditions. Think not meanly of the serpent, "it is the very omnipotence of the earth. That rivulet of smooth silver—how does it flow, think you? It literally rows on the earth with every scale for an oar; it bites the dust with the ridges of its body. Watch it when it moves slowly— a wave, but without wind! a current, but with no fall! all the body moving at the same instant, yet some of it to one side, some to another, or some forward, and the rest of the coil backwards; but all with the same calm will and equal way—no contraction, no extension; one soundless, causeless, march of sequent rings, and spectral procession of spotted dust, with dissolution in its fangs, dislocation in its coils. Startle it—the winding stream will become a twisted arrow; the wave of poisoned life will lash through the grass like a cast lance. It scarcely breathes with its one lung (the other shrivelled and abortive); it is passive to the sun and shade, and cold or hot like a stone; yet 'it can outclimb the monkey, outswim the fish, outleap the zebra, outwrestle the athlete, and crush the tiger.' It is a Divine hieroglyph of the demoniac power of the earth—of the entire earthly nature. As the bird is the clothed power of the air, so this is the clothed power of the dust; as the bird is the symbol of the spirit of life, so this of the grasp and sting of death."[1]

This well-known and eloquent passage is not perfectly true,— thus the serpent breathes not scarcely but strongly with its one lung,—but, while you may correct and complete it as you will, I am sure that you will find here more insight into the nature of serpents than in pages of anatomical description.

---

[1] Ruskin's *Queen of the Air*.

A few snakes have mouths which do not distend, skull bones which are slightly movable, teeth in one jaw (upper or lower) only, and rudiments of hind legs. These are included in the genera *Typhlops* and *Anomalepsis*, and are small simple ophidians.

Many are likewise non-venomous snakes, but with wider gape and more mobile skull bones, and with simple teeth on both jaws. Some are very large and have great powers of strangling. Such are the Pythons, the Boa, and the Anaconda. To these our grass snake (*Tropidonotus-natrix*) is allied.

Many poisonous snakes have large permanently erect grooved fangs in the upper jaw, and a salivary gland whose secretion is venomous. Such are the cobra (*Naja tripudians*), the Egyptian asp (*Naja haje*), the coral snakes (*Elaps*), and the sea snakes (*Hydrophis*).

Other poisonous snakes have perforated fang teeth, which can be raised and depressed. Such are the vipers (*Vipera*), the British adder (*Pelias berus*), the copperhead (*Ancistrodon contortrix*), the rattlesnakes (*Crotalus*).

Tortoises and Turtles (Chelonia).—Boxed in by a bony shield above and by a bony shield below, and often with partially retractile head and tail and legs, the Chelonians are thoroughly armoured. On the average the pitch of their life is low, but their tenacity of life is great. Slow in growth, slow in movement, slow even in reproduction are many of them, and they can endure long fasting. It is said that a tortoise walked at least 200 yards, twenty-four hours after it was decapitated, while it is well known that the heart of a tortoise will beat for two or three days after it has been isolated from the animal. In connection with their sluggishness it is significant that the ribs which help to some extent in the respiratory movements of higher animals are soldered into the dorsal shield, thus sluggish respiration may be in part the cause, as it is in part the result, of constitutional passivity. All the Chelonians lay eggs in nests scooped in the earth or sand.

The marine turtles (*e.g. Sphargis, Chelone*), the estuarine soft-shelled turtles (*e.g. Aspidonectes*), the freshwater turtles (*e.g. Emys*), and the snapping turtle (*Chelydra*) are more active than the land tortoises, such as the European *Testudo græca*, often kept as a pet. The tortoise of the Galapagos Islands (*Testudo elephantopus*), the river tortoise (*Podocnemys expansa*) of the Amazon, the bearded South American turtle (*Chelys matamata*), and the green turtle (*Chelone mydas*) attain a large size, sometimes measuring about 3 feet in length.

Crocodilians (Crocodilia).—Crocodiles, alligators, and gavials seem in our present perspective very much alike—strong, large, heavily armoured reptiles, at home in tropical rivers, but clumsy and stiff-necked on land, feeding on fishes and small mammals, growing slowly and without that definite limit which punctuates the life-history of most animals, attaining, moreover, a great age, freed after youth is past from the attacks of almost every foe but man. The teeth are firmly implanted in sockets; the limbs and tail are suited for swimming, and also for crawling; the heart is more highly developed than in other reptiles, having four instead of three chambers. The animals lie in wait for victims, and usually drown them, being themselves able to breathe while the mouth is full of water, if only the nostrils be kept above the surface.

In many ways Reptiles touch human life, the poisonous snakes are very fatal, especially in India; crocodilians are sometimes destructive; turtles afford food and "tortoise shell;" lizards are delightfully beautiful.

8. **Birds.**—What mammals are to the earth, and fishes to the sea, birds are to the air. Has anything truer ever been said of

FIG. 55.—The Collocalia, which from the secreted juice of its salivary glands builds the edible-bird's-nest. (Adapted from Brehm.)

them than this sentence from Ruskin's *Queen of the Air*? "The bird is little more than a drift of the air brought into form by plumes; the air is in all its quills, it breathes through its whole frame and flesh, and glows with air in its flying, like a blown

flame : it rests upon the air, subdues it, surpasses it, outraces it ;—
is the air, conscious of itself, conquering itself, ruling itself."

Birds represent among animals the climax of activity, an index to which may be found in their high temperature, from 2°-14° Fahrenheit higher than that of mammals. In many other ways they rank high, for whether we consider the muscles which move the wings in flight, the skeleton which so marvellously combines strength with lightness, the breathing powers perfected and economised by a set of balloons around the lungs, or the heart which drives and receives the warm blood, we recognise that birds share with mammals the position of the highest animals. And while it is true that the brains of birds are not wrinkled with thought like those of mammals, and that the close connection between mother and offspring characteristic of most mammals is absent in birds, it may be urged by those who know their joyousness that birds feel more if they think less, while the patience and solicitude connected with nest-making and brooding testify to the strength of their parental love. Usually living in varied and beautiful surroundings, birds have keen eyes and sharp ears, tutored to a sense of beauty, as we may surely conclude from their cradles and love songs. They love much and joyously, and live a life remarkably free and restless, qualities symbolised by the voice of the air in their throat, and by the sunshine of their plumes. There is more than zoological truth in saying that in the bird "the breath or spirit is more full than in any other creature, and the earth power least," or in thinking of birds as the purest embodiments of Athene of the air.

But just as there are among mammals feverish bats with the power of true flight, and whales somewhat fish-like, so there are exceptional birds, runners like the ostriches and cassowaries, swimmers like the penguins, criminals too like the cuckoos and cow-birds in which the maternal instincts are strangely perverted. As we go back into the past, strange forms are discovered, with teeth, long tails, and other characteristics which link the birds of the air to the grovelling reptiles of the earth. Even to-day there lives a "reptilian-bird"—*Opisthocomus*—which has retained more than any other indisputable affinities with the reptiles. Professor W. K. Parker, one of the profoundest of all students of birds, described this form in one of his last papers, and there used a comparison which helps us to appreciate birds. They are among backboned animals what insects are among the backboneless—winged possessors of the air, and just as many insects pass through a caterpillar and chrysalis stage before reaching the acme of their life as a flying imago, so do the young birds within the veil of the eggshell pass through somewhat fish-like and somewhat reptile-like

Fig. 56.—Decorative male and less adorned female of Spathura—a genus of Humming-birds. (From Darwin, after Brehm.)

CHAP. XVI   *Backboned Animals*   267

stages before they attain to the possession of wings and the enjoyment of freedom.

The great majority of birds are fliers, and possess a keeled breast-bone, to which are fixed the muscles used in flight. To this keel or carina they owe their name Carinatæ. The flying host includes the gulls and grebes, the plovers and cranes, the ducks and geese, the storks and herons, the pelicans and cormorants, the partridges and pheasants, the sand grouse, the pigeons, the birds of prey, the parrots, the pies, and about 6000 Passerine or sparrow-like birds, including thrushes and warblers, wrens and swallows, finches and crows, starlings and birds of paradise. To these orders we have to add *Opisthocomus*, from which it is perhaps easier to pass to some of the keeled fossil birds, some of which possessed teeth.

Distinct from the keeled fliers, both ancient and modern, are the running-birds, which are incapable of flight, and therefore possess a flat raft-like breast bone, to which they owe their title Ratitæ. Nowadays these are few in number, the Ostrich and the Rhea, the Cassowary and Emu, and the small Kiwi. Beside these must be ranked the giant Moa of New Zealand, not long extinct, and the more ancient, not less gigantic *Æpyornis* of Madagascar, while farther back still, from the Chalk strata of America, the remains of toothed keelless birds have been disentombed.

The most reptilian, least bird-like of birds is the oldest fossil of all, placed in a sub-class by itself, the *Archæopteryx* (lit. ancient bird) from strata of Jurassic age.

FIG. 57.—Restoration of the extinct moa (*Dinornis ingens*), and alongside of it the little kiwi (*Apteryx mantelli*). (From Chambers's *Encyclop.*; after F. v. Hochstetter.)

9. **Mammalia.**—Of the highest class of animals—the Mammalia—I need say least for they are most familiar. Most of them are terrestrial, four-footed, and hairy. Bats and whales, seals and sea-cows, are obviously excep-

tional. The brain of mammals is more highly developed than that of other animals, and in the great majority there is a prolonged (placental) connection between the unborn young and the mother. In all cases the mothers feed the tender young with milk.

In the class there are three grades:—

(1) In the Duckmole (*Ornithorhynchus*) and the Porcupine Ant-Eater (*Echidna*), and perhaps another genus *Proechidna*, the females lay eggs. In many other ways these exclusively Australasian mammals are primitive, exhibiting affinities with reptiles.

(2) In the Marsupials, which, with the exception of some American Opossums, are also Australasian, the young are born at a very tender age, as it were, prematurely. In the great majority of genera, the mothers stow them away in an external pouch, where they are fed and sheltered till able to fend for themselves. In Australia the Marsupials have been saved by insulation from stronger mammals, which seem to have exterminated them in other parts of the earth, the Opossums which hide in American forests being the only Marsupials surviving outside Australasia, though fossils show that the race had once a much wider distribution. In their Australian retreat, apart from all higher Mammalia (mice, rabbits, and the like being modern imports) the Marsupials have evolved along many lines, prophetic of the higher orders of mammals. There are "carnivores" like the Thylacine and the Dasyure, "herbivores" like the Kangaroos, "insectivores" like the banded ant-eater *Myrmecobius*, and "rodents" like the Wombat.

(3) In all the other orders of mammals there is a close connection between mother and unborn offspring.

Two orders are lowly and distinctly separate from the others and from one another—the Edentata represented by sloths, ant-eaters, armadillos, pangolins, and the Aard-Vark; and the Sirenia or Sea-Cows which now include only the dugong and the manatee.

Along one fairly definite line we may rank three other orders —the Insectivores, the Bats, and the Carnivores. The hedgehog, which is at once a lowly and a central type of mammal, may be taken as the beginning of this line. Along with shrews, moles, porcupines, the hedgehogs form the order Insectivora. To these the Bats (Cheiroptera), with their bird-like powers of flight, are linked, while the Carnivora (cats, dogs, bears, and seals), though progressive in a different direction, seem also related.

Comparable to the Insectivores, but on a different line, are the gnawing Rodents, rabbits and hares, rats and mice, squirrels and beavers. This line leads on to the Elephants, from the company of which the mammoths have disappeared since man arose on the earth. With the Elephants, the rock-coneys or Hyraxes—"a feeble folk"—seem to be allied. Both are often included in the great order of hoofed animals or Ungulates, along with the odd-toed

FIG. 58.—*Phenacodus primævus*, a primitive extinct mammal from the lower Eocene of N. America. The actual size of the slab of rock on which it rested was 49 inches in length. (From Chambers's *Encyclop.*; after Cope.)

animals—horse, rhinoceros, and tapir, and a larger number of even-toed forms, hog and hippopotamus, camel and dromedary,

FIG. 59.—Head of gorilla. (From Du Chaillu.)

and the true cud-chewers or ruminants such as sheep and cattle, deer and antelopes. From the ancient predecessors of the modern

Ungulates, it seems likely enough that the Cetaceans (whales and dolphins) diverged.

A third line, which we may call median, leads through the Lemurs on to Monkeys. It must be noted, however, that these lines, which seem distinct from one another if we confine our attention to living mammals, are linked by extinct forms. Thus a

FIG. 60.—Head of male Semnopithecus. (From Darwin.)

remarkable fossil type, *Phenacodus*, is regarded by Cope as presenting affinities with Ungulates, Lemurs, and Carnivores.

The monkeys which most closely resemble man in structure, habits, and intelligence, are the so-called anthropoid apes, the gorilla, the chimpanzee, the orang-utan, and the gibbon. A second grade is represented by the more dog-like, narrow-nosed Old World apes, such as the baboons and mandrills. Lower in many ways are the broad-nosed New World or American monkeys, *e.g.* the numerous species of *Cebus*, some of which are the familiar companions of itinerant musicians, while lowest and smallest among true monkeys are the South American marmosets. Distinct from all these, probably outside the monkey order altogether, are the so-called half-monkeys or Lemurs.

We might describe the clever activities of monkeys, the shelters which some of them make, their family life, parental care and sociality, their docility, their intelligent habits of investigation, and their quickness to profit by experience ; but it would all amount to this, that their life at many points touches the human, that they are in some ways like growing children, in other ways like savage men, though with more circumscribed limits of progress than either.

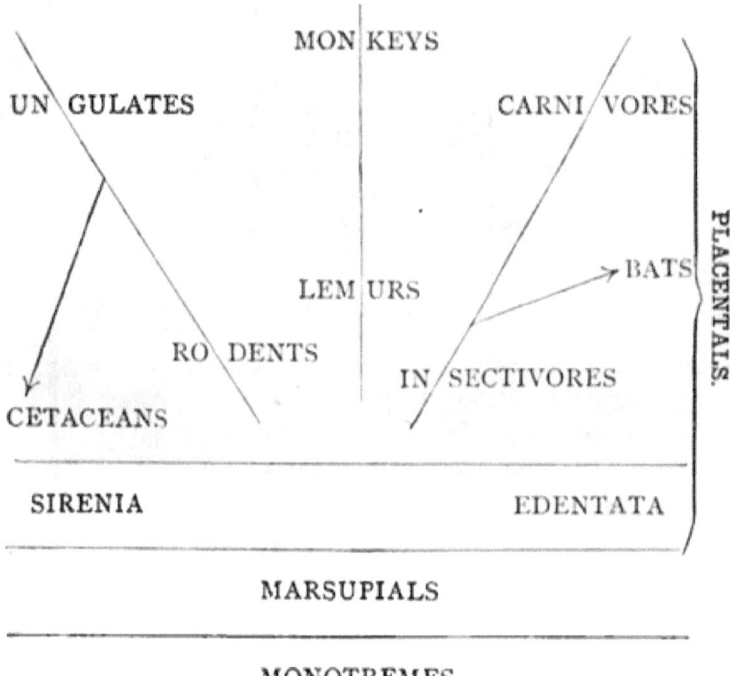

ORDERS OF MAMMALS.

SURVEY OF THE ANIMAL KINGDOM

VERTEBRATES.	**CŒLOMATA.**	BIRDS. Flying-Birds. Running-Birds.	MAMMALS.	Placentals. Marsupials. Monotremes.
		Snakes. Lizards. REPTILES. Crocodiles. Tortoises.		
		FISHES. Double-Breathers. Bony-Fishes. Ganoids. Elasmobranchs.	AMPHIBIANS. Newt. Frog. CYCLOSTOMATA. Lamprey. Hagfish.	
		LANCELET.	TUNICATES.	

METAZOA.

INVERTEBRATES.

Insects. Arachnids. Myriapods. Peripatus. ARTHROPODS. Crustaceans.	BALANOGLOSSUS. ANNELIDS. "WORMS." FLAT-WORMS.	Cuttlefish. Gasteropods. MOLLUSCS. Bivalves. Feather-stars. Brittle-stars. Starfish. ECHINODERMS. Sea-urchins. Sea-cucumbers.

Ctenophores. Jellyfish. Sea-Anemones. Corals.
STINGING-ANIMALS or CŒLENTERATES.
Medusoids and Hydroids.

SPONGES.

Infusorians. Rhizopods. Gregarines.
SIMPLEST ANIMALS.

PROTOZOA.

PART IV

THE EVOLUTION OF ANIMAL LIFE

CHAPTER XVII

THE EVIDENCES OF EVOLUTION

1. *The Idea of Evolution*—2. *Arguments for Evolution: Physiological, Morphological, Historical*—3. *Origin of Life*

WE observe animals in their native haunts, and study their growth, their maturity, their loves, their struggles, and their death; we collect, name, preserve, and classify them; we cut them to pieces, and know their organs, tissues, and cells; we go back upon their life and inquire into the secret working of their vital mechanism; we ransack the rocks for the remains of those animals which lived ages ago upon the earth; we watch how the chick is formed within the egg, and yet we are not satisfied. We seem to hear snatches of music which we cannot combine. We seek some unifying idea, some conception of the manner in which the world of life has become what it is.

1. **The Idea of Evolution.**—We do not dream now, as men dreamed once, that all has been as it is since all emerged from the mist of an unthinkable beginning; nor can we believe now, as men believed once, that all came into its present state of being by a flash of almighty volition. We still dream, indeed, of an unthinkable beginning, but we know that the past has been full of change; we still

T

believe in almighty volition, but rather as a continuous reality than as expressed in any event of the past. Thus Erasmus Darwin (1794), speaking of Hume, says "he concluded that the world itself might have been generated rather than created; that it might have been gradually produced from very small beginnings, increasing by the activity of its inherent principles, rather than by a sudden evolution of the whole by the Almighty fiat." In short, we have extended to the world around us our own characteristic perception of human *history*; we have concluded that in all things the present is the child of the past and the parent of the future.

But while we dismiss the theory of permanence as demonstrably false, and the theory of successive cataclysms and re-creations as improbable,[1] without feeling it necessary to discuss either the falsity or the improbability, we must state on what basis our conviction of continuous evolution rests. "La nature ne nous offre le spectacle d'aucune creation, c'est d'une continuation éternelle." "As in the development of a fugue," Samuel Butler says, "where, when the subject and counter-subject have been announced, there must thenceforth be nothing new, and yet all must be new, so throughout organic nature—which is a fugue developed to great length from a very simple subject—everything is linked on to and grows out of that which comes next to it in order—errors and omissions excepted."

2. **Arguments for Evolution.**—What then are the facts which have convinced naturalists that the plants and the animals of to-day are descended from others of a simpler sort, and the latter from yet simpler ancestors, and so on, back and back to those first forms in which all that succeeded were implied? I refer you to Darwin's *Origin of Species* (1859), where the arguments were marshalled in such a masterly fashion that they forced the conviction

---

[1] I use the word in its literal sense—"not admitting of proof." It is not my duty nor my desire to discuss the poetical, or philosophical, or religious conceptions which lie behind the concrete cosmogonies of different ages and minds. To many modern theologians creation really means the institution of the order of nature, the possibility of natural evolution included.

of the world. To the statements of the case by Spencer, Haeckel, Huxley, Romanes, and others, I have given references in the chapter on books. Darwin's arguments were derived (*a*) from the distribution of animals in space; (*b*) from their successive appearance in time, (*c*) from actual variations observed in domestication, cultivation, and in nature; (*d*) from facts of structure, *e.g.* homologous and rudimentary organs, (*e*) from embryology. I shall simply illustrate the different kinds of evidence, and that under three heads—(*a*) physiological, (*b*) structural, (*c*) historical.

(*a*) **Physiological.**—A study of the life of organisms shows that the ancient and even Linnæan dogma of the constancy or immutability of species was false. Organisms change under our eyes. They are not like cast-iron; they are plastic. One of the most striking cases in the Natural History Collection of the British Museum is that near the entrance, where on a tree are perched domesticated pigeons of many sorts—fantail, pouter, tumbler, and the like—while in the centre is the ancestral rock-dove *Columba livia*, from which we know that all the rest have been derived. In other domesticated animals, even when we allow that some of them have had multiple origins, we find abundant proof of variability. But what occurs under man's supervision in the domestication of animals and in the cultivation of plants occurs also in the state of nature. Natural "varieties" which link species to species are very common, and the offspring of one brood differ from one another and from their parents. How many strange sports there are and grim reversions! and, as we shall afterwards see, modifications of individuals by force of external conditions are not uncommon. Those who say they see no variation now going on in nature should try a month's work at identifying species. I have known of an ancient man who dwelt in a small town; he did not believe in the reality of railways and to him the testimony of observers was as an idle tale; he was not daunted in his scepticism even when the railway was extended to his town, for he was aged, and remained at home, dying a professed unbeliever in that which he had

FIG. 61.—Varieties of domestic pigeon arranged around the ancestral rock-dove (*Columba livia*). (Based on Darwin's figures.)

never seen. Conviction depends on more than intelligence, often on emotional vested interests.

(*b*) **Morphological.**—There are said to be over a million species of living animals, about half of them insects. Even their number might suggest blood-relationship, but our recognition of this becomes clear when we see that species is often united to species, genus to genus, and even class to class, by connecting links. The fact that we can make at least a plausible genealogical tree of animals, arranging them in series along the lines of hypothetical pedigree, is also suggestive.

Throughout long series, structures fundamentally the same appear with varied form and function; the same bones and muscles are twisted into a variety of shapes. Why this adherence to type if animals are independent of one another? How necessary it is if all are branches of one tree.

By rudimentary organs also the same conclusion is suggested. What mean the unused gill-clefts of reptiles, birds, and mammals, unless the ancestors of these classes were fish-like; what mean the teeth of very young whalebone whales, of an embryonic parrot and turtle, unless they are vestiges of those which their ancestors possessed? There are similar vestigial structures among most animals. In man alone there are about seventy little things which might be termed rudimentary; his body is a museum of relics. We are familiar with unsounded or rudimentary letters in many words; we do not sound the "o" in leopard nor the "l" in alms, but from these rudimentary letters we read the history of the words.

(*c*) **Historical.**—Every one recognises that animals have not always been as they now are; we have only to dig to be convinced that the fauna of the earth has had a history. But it does not follow that the succession of fauna after fauna, age after age, has been a progressive development. What evidence is there of this?

In the first place, there is the general fact that fishes appear before amphibians, and these before reptiles, and these before birds, and that the same correspondence between order of appearance and structural rank is often true in detail within the separate classes of animals. There are some marvellously complete series of fossils, especially, perhaps, that of the extinct cuttlefishes, in which the steps of progressive evolution are still traceable. Moreover, the long pedigree of some animals, such as the horse, has been worked out so perfectly that more convincing demonstration is hardly possible. In Professor Huxley's *American Addresses*, or in that pleasant introduction to zoology afforded by Professor W. H. Flower's little book on the horse (Modern Science Series, Lond., 1891), you will find the story of the horse's pedigree most lucidly told: how in early Eocene times there lived small quadrupeds about the size of sheep that walked securely upon five toes, how these animals lost, first the inner toe, while the third grew larger, and then the fifth; how the third continued to grow larger and the second and fourth to become smaller until they disappeared almost entirely, remaining only as small splint bones; and how thus the light-footed runners on tiptoe of the dry plains were evolved from the short-legged splay-footed plodders of the Eocene marshes. Finally, there are many extinct types which link order to order and even class to class, such as that strange mammal *Phenacodus*, which seems to occupy a central position in the series, so numerous are its affinities, or such as those saurians which link crawling reptile to soaring bird.

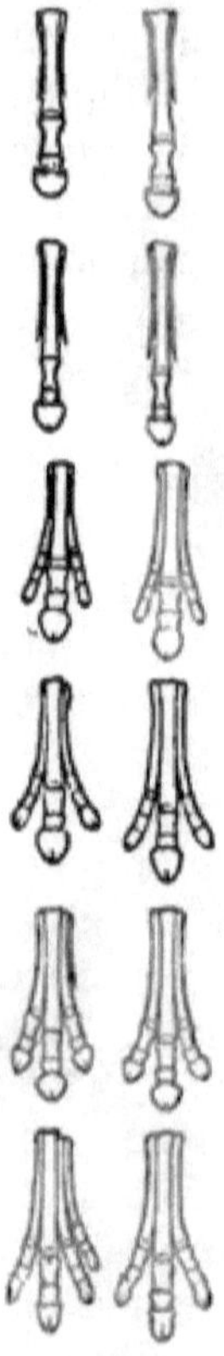

FIG. 62.—Fore and hind feet of the horse and some of its ancestors, showing the gradual reduction in the number of digits. (From Chambers's *Encyclop.*; after Marsh.)

Another historical argument of great importance is that derived from the study of the geographical distribution of animals, but this cannot be appreciated without studying the detailed facts. These suggest that the various types of animals have spread from definite centres, along convenient paths of diffusion, varying into species after species as their range extended.

But the history of the individual is even more instructive. The first three grades of structure observed among living animals are: (1) Single cells (most Protozoa), (2) balls of cells (a few Protozoa which form colonies), and (3), two-layered sacs of cells (*e.g.* the simplest sponges). But these three grades correspond to the first three steps in the individual life-history of any many-celled animal. Every one begins as a single cell, at the presumed beginning again; this divides into a ball of cells, the second grade of structure; the ball becomes a two-layered sac of cells. The

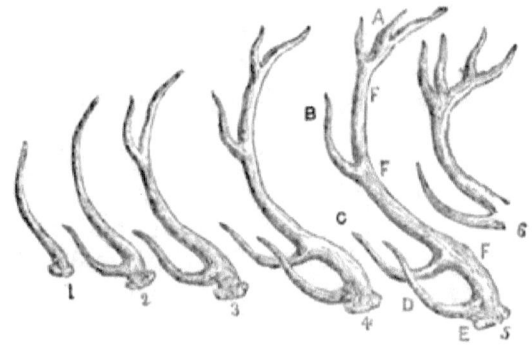

FIG. 63.—Antlers of deer (1-5) in successive years; but the figure might almost represent at the same time the degree of evolution exhibited by the antlers of deer in successive ages. (From Chambers's *Encyclop.*)

correspondence between the first three grades of structure and the first three chapters in the individual's life-history is complete. It is true as a general statement that the individual development proceeds step by step along a path approximately parallel to the presumed progress of the race, so far as that is traceable from the successive grades of structure and from the records of the rocks. Even in regard to details such as the development of antlers on stags the parallelism of racial and individual history may be observed. Of this correspondence it is difficult to see any elucidation except that the individual in its life-history in great part re-treads the path of ancestral evolution.

I have illustrated these evidences of evolution very

briefly, for they have been stated many times of late years. The idea of evolution has also justified itself by the light which it has cast not only on biological, but on physical, psychological, and sociological facts. There has never been a more germinal idea; it is fast becoming organic in all our thinking.

To those who feel a repugnance to the doctrine of descent, I suggest the following considerations:—

(1) In so far as conclusions do not affect conduct, it seems wise to conserve what makes one happiest. If your intellectual and emotional necessities are better satisfied, for instance, by any one of the creationist theories than by that of a gradual and natural progress from simple beginnings to implied ends, and if you feel that your sense of the marvel, beauty, and sacredness of life would be impoverished by a change of theory, then I should not seek to persuade you.

(2) But as we do not think a tree less stately because we know the tiny seed from which it grew, nor any man less noble because he was once a little child, so we ought not to look on the world of life with eyes less full of wonder or reverence, even if we feel that we know something of its humble origins.

(3) Finally, we should be careful to distinguish between the doctrine of natural descent, which, to most naturalists, seems a solemn fact, and the theories of evolution which explain how the progressive descent was brought about. For in regard to the causal, as distinguished from the modal explanation of the world, we are or ought to be uncertain.

3. **Origin of Life.**—It is no dogma, nor yet a "law of Biogenesis," but a fact of experience, to which no exception has been demonstrated, that living organisms arise from pre-existent organisms—*Omne vivum e vivo.*

As to the origin of life upon the earth we know nothing, but hold various opinions. (1) Thus it is believed that life began independently of those natural conditions which come within the ken of scientific inquirers; in other words, it is believed that the first living things were created. That this belief presents intellectual difficulties to many minds may mean that its fittest expression in words has not been attained, or is unattainable. (2) It has been suggested that germs of life reached this earth in the bosom of meteorites from somewhere else. This at least shifts the responsibility of the problem off the shoulders of this planet. (3) It is suggested that living matter may have been evolved from not-living matter on the earth's surface. If we accept this suggestion, we must of course suppose that in not-living matter the qualities characteristic of living organisms are implicit. The evolutionist's common denominator is then as inexpressibly marvellous as the philosopher's greatest common measure.

CHAPTER XVIII

THE EVOLUTION OF EVOLUTION THEORIES

1. *Greek Philosophers*—2. *Aristotle*—3. *Lucretius*—4. *Evolutionists before Darwin*—5. *Three old Masters: Buffon, Erasmus Darwin, Lamarck*—6. *Charles Darwin*—7. *Darwin's Fellow-workers*—8. *The Present State of Opinion*

THE conception of evolution is no new idea, it is the human idea of history grown larger, large enough to cover the whole world. The extension of the idea was gradual, as men felt the need of extending it; and at the same moment we find men believing in the external permanence of one set of phenomena, in the creation of others, in the evolution of others. One authority says human institutions have been evolved; man was created; the heavens are eternal. According to another, matter and motion are eternal; life was created; the rest has been evolved, except, perhaps, the evolution theory which was created by Darwin.

1. **Greek Philosophers.**—Of the wise men of Greece and what they thought of the nature and origin of things, I shall say little, for I have no direct acquaintance with the writings of those who lived before Aristotle. Moreover, though an authority so competent as Zeller has written on the "Grecian predecessors of Darwin," most of them were philosophers not naturalists, and we are apt to read our own ideas into their words. They thought, indeed, as we are thinking, about the physical and organic universe, and some of them believed it to be, as we do, the result of

a process; but here in most cases ends the resemblance between their thought and ours.

Thus when Anaximander spoke of a fish-like stage in the past history of man, this was no prophecy of the modern idea that a fish-like form was one of the far-off ancestors of backboned animals, it was only a fancy invented to get over a difficulty connected with the infancy of the first human being.

Or, when we read that several of these sages reduced the world to one element, the ether, we do the progress of knowledge injustice if we say that men are simply returning to this after more than two thousand years. For that conception of the ether which is characteristic of modern physical science has been, or is being, slowly attained by precise and patient analysis, whereas the ancient conception was reached by metaphysical speculation. If we are returning to the Greeks, it is on a higher turn of the spiral, so far at least as the ether is concerned.

When we read that Empedocles sought to explain the world as the result of two principles—love and hate—working on the four elements, we may, if we are so inclined, call these principles "attractive and repulsive forces"; we may recognise in them the altruistic and individualistic factors in organic evolution, and what not; but Empedocles was a poetic philosopher, no far-sighted prophet of evolution.

But the student cannot afford to overlook the lesson which Democritus first clearly taught, that we do not explain any result until we find out the natural conditions which bring it about, that we only understand an effect when we are able to analyse its causes. We require a so-called "mechanical," or more strictly, a dynamical explanation of results. It is easy to show that it is advantageous for a root to have a root-cap, but we wish to know how the cap comes to be there. It is obvious that the antlers of a stag are useful weapons, but we must inquire as precisely as possible how they first appeared and still grow.

2. **Aristotle.**—As in other departments of knowledge, so in zoology the work of Aristotle is fundamental. It is

wonderful to think of his knowledge of the forms and ways of life, or the insight with which he foresaw such useful distinctions as that between analogous and homologous organs, or his recognition of the fact of correlation, of the advantages of division of labour within organisms, of the gradual differentiation observed in development. He planted seeds which grew after long sleep into comparative anatomy and classification. Yet with what sublime humility he says: " I found no basis prepared, no models to copy. Mine is the first step, and therefore a small one, though worked out with much thought and hard labour." Aristotle was not an evolutionist, for, although he recognised the changefulness of life, the world was to him an eternal fact not a stage in a process.

" In nature, the passage from inanimate things to animals is so gradual that it is impossible to draw a hard-and-fast line between them. After inanimate things come plants, which differ from one another in the degree of life which they possess. Compared with inert bodies, plants seem endowed with life; compared with animals, they seem inanimate. From plants to animals the passage is by no means sudden or abrupt; one finds living things in the sea about which there is doubt whether they be animals or plants."
" Animals are at war with one another when they live in the same place and use the same food. If the food be not sufficiently abundant they fight for it even with those of the same kind."

3. **Lucretius.**—Among the Romans Lucretius gave noble expression to the philosophy of Epicurus. I shall not try to explain his materialistic theory of the concourse of atoms into stable and well-adapted forms, but rather quote a few sentences in which he states his belief that the earth is the mother of all life, and that animals work out their destiny in a struggle for existence. He was a cosmic, but hardly an organic evolutionist, for, according to his poetic fancy, organisms arose from the earth's fertile bosom and not by the gradual transformation of simpler predecessors.

" In the beginning the earth gave forth all kinds of herbage and verdant sheen about the hills and over all the plains; the flowery meadows glittered with the bright green hue, and next in order to the different trees was given a strong and emulous desire of growing up into the air with full unbridled powers. . . . With good reason the earth has gotten the name of mother, since all things have been produced out of the earth. . . .

"We see that many conditions must meet together in things in order that they may beget and continue their kinds; first a supply of food, then a way in which the birth-producing seeds throughout the frame may stream from the relaxed limbs. . . . And many races of living things must then have died out and been unable to beget and continue their breed. For in the case of all things which you see breathing the breath of life, either craft or courage or else speed has from the beginning of its existence protected and preserved each particular race. And there are many things which, recommended to us by their useful services, continue to exist consigned to our protection.

"In the first place, the first breed of lions and the savage races their courage has protected, foxes their craft, and stags their proneness to flight. But light-sleeping dogs with faithful heart in breast, and every kind which is born of the seed of beasts of burden, and at the same time the woolly flocks and the horned herds, are all consigned to the protection of man. For they have ever fled with eagerness from wild beasts, and have ensued peace, and plenty of food obtained without their own labour, as we give it in requital of their useful services. But those to whom nature has granted none of these qualities, so that they could neither live by their own means nor perform for us any useful service, in return for which we should suffer their kind to feed and be safe under our protection, those, you are to know, would lie exposed as a prey and booty of others, hampered all in their own death-bringing shackles, until nature brought that kind to utter destruction."

4. **Evolutionists before Darwin.**—From Lucretius I shall pass to Buffon, for the intervening centuries were uneventful as regards zoology. Hugo Spitzer, one of the historians of evolution, finds analogies between certain mediæval scholastics and the Darwinians of the nineteenth century, but these are subtle comparisons. Yet long before Darwin's day there were evolutionists, and the first of these who can be called great was Buffon.

We must guard against supposing that the works of Buffon, or Lamarck, or Darwin were inexplicable creations of genius, or that they came like cataclysms, without warning, to shatter the conventional traditions of their time. For all great workers have their forerunners, who prepare their paths. Therefore in thinking out the history of evolutionist theories before that of Buffon, we must take account of many forces which began to be influential from the twelfth century onwards. "Evolution in social affairs has not only suggested our ideas of evolution in the other sciences, but has deeply coloured them in accordance with the particular phase of social evolution current at the time."[1] In other words, we must abandon the idea that we can understand the history of any science as such, without reference to contemporary evolution in other departments of activity. The evolution of theories of evolution is bound up with the whole progress of the world.

In trying to determine those social and intellectual forces of which the modern conception of organic evolution has been a resultant, we should take account of social changes, such as the collapse of the feudal system, the crusades, the invention of printing, the discovery of America, the French Revolution, the beginning of the steam age; of theological and religious movements, such as the Protestant Reformation and the spread of Deism; of a long series of evolutionist philosophers, some of whom were at the same time students of the physical sciences, — notably Descartes, Spinoza, Leibnitz, Herder, Kant, and Schelling; of the acceptance of evolutionary conceptions in regard to other orders of facts, especially in regard to the earth and the solar system; and, finally, of those few naturalists, like De Maillet and Robinet, who, before Buffon's day, whispered evolutionist heresies. (The history of an idea is like that of an organism in which cross-fertilisation and composite inheritance complicate the pedigree.)

5. **Three old Masters.**—Among the evolutionists before Darwin I shall speak of only three—Buffon, Erasmus Darwin, and Lamarck.

BUFFON (1707-1788) was born to wealth and was wedded to Fortune. He sat in kings' houses, his statue adorned their gardens. As Director of the *Jardin du Roi* he had opportunity to acquire a wide knowledge of animals. He commanded the assistance of able collaborateurs, and his own

[1] Article "Evolution" (P. Geddes) in Chambers's *Encyclopædia*.

industry was untiring. He was about forty years old when he began his great Natural History, and he worked till he was fourscore. He lived a full life, the success of which we can almost read in the strong confidence of his style. 'Le style, c'est l'homme même," he said ; or again, " Le style est comme le bonheur ; il vient de la douceur de l'âme." Rousseau called him "La plus belle plume du siècle;" Mirabeau said, "Le plus grand homme de son siècle et de bien d'autres ;" Voltaire first mocked and then praised him ; and Diderot also eulogised. Buffon was first a man then a zoologist, which seems to be the natural, though by no means universally recognised, order of precedence, and we have pleasant pictures of his handsome person, his magnificence, his diplomatic manners, and a splendid genius, which he himself called "a supreme capacity for taking pains."

Buffon's culture was very wide. He had an early training in mathematics, and translated Newton's *Fluxions*; he seems to have been familiar with the chemistry and physics of his time ; he was curious about everything. Before Laplace, he elaborated an hypothesis as to the origin of the solar system ; before Hutton and Lyell, he realised that causes like those now at work had in the long past sculptured the earth ; he had a special theory of heredity not unlike Darwin's, and a by no means narrow theory of evolution, in which he recognised the struggle for existence and the elimination of the unfit, the influence of isolation and of artificial selection, but especially the direct action of food, climate, and other surrounding influences upon the organism. It is generally allowed that there is in Buffon's writings something of that indefiniteness which often characterises pioneer works, and a lack of depth not unnatural in a survey so broad, but they exhibit some remarkable illustrations of prophetic genius, and a lively appreciation of nature.

It is probable that Buffon's treatment of zoology gained freedom because he wrote in French, having shaken off the shackles which the prevalent custom of writing in Latin imposed, and it cannot be doubted that his works did something to prepare the way for the future reception of the doctrine of descent. He had a vivid feeling of the unity of nature, throwing out hints in regard to the fundamental similarity of different forms of matter, suggesting that heat and light are atomic **movements,** denying the existence of hard-and-fast lines—" Le vivant et l'animé est une propriété physique de la matière!" protesting against crude distinctions between **plants** and **animals, and realising** above all that **there is** one great family of life. Naturalists **had** been wandering up and down the valleys studying **their** characteristic contours; Buffon took an eagle's flight **and saw the** connected range of hills,—" l'enchainement des êtres."

ERASMUS DARWIN (1731-1802), grandfather to the author of the *Origin of Species*, was a large-hearted, thoughtful physician, whose life was as full of pleasant eccentricities, as his stammering speech of wit, and his books of wisdom. We have pleasant pictures of the philosophical physician of Lichfield and Derby, driving about in a whimsical **un**stable carriage of his own contrivance, prescribing abundant food and cowslip wine, rich in good health and generosity. Comparing his writings with those of Buffon, an acquaintance with which he evidently possessed, we find more emotion and intensity, more of the poet and none of the diplomatist. He approached the study of organic life on the one hand as a physician and physiologist, on the other hand as a gardener and lover of plants; and, apart from poetic conceits, his writings are characterised by a directness and simplicity of treatment which we often describe as " common-sense."

He believed that the different **kinds of** plants and animals were descended **from a** few ancestral forms, or possibly from **one and** the same kind **of** "vital filament," and that evolutionary change **was** mainly due to the exertions which organisms made to preserve or better themselves. He showed that animals were driven to exertion by hunger, by love, and by the need of protection, and explained their progress as the result of their endeavours. Buffon underrated the transforming influence of action, and laid emphasis upon the direct influence of surroundings; Erasmus Darwin emphasised function, **and** regarded the influence of the

The Evolution of Evolution Theories

environment as for the most part indirect. Let us quote some conclusions from his *Zoonomia* (1794):—

"Owing to the imperfection of language the offspring is termed a *new* animal, but is in truth a branch or elongation of the parent, since a part of the embyron animal is, or was, a part of the parent, and therefore in strict language cannot be said to be entirely *new* at the time of its production; and therefore it may retain some of the habits of the parent-system."

"The fetus or embryon is formed by apposition of new parts, and not by the distention of a primordial nest of germs included one within another like the cups of a conjuror."

"From their first rudiment, or primordium, to the termination of their lives, all animals undergo perpetual transformations; which are in part produced by their own exertions in consequence of their desires and aversions, of their pleasures and their pains, or of irritations, or of associations; and many of these acquired forms or propensities are transmitted to their posterity."

"As air and water are supplied to animals in sufficient profusion, the three great objects of desire, which have changed the forms of many animals by their exertions to gratify them, are those of lust, hunger, and security."

"This idea of the gradual generation of all things seems to have been as familiar to the ancient philosophers as to the modern ones, and to have given rise to the beautiful hieroglyphic figure of the πρῶτον ᾠόν, or first great egg, produced by night, that is, whose origin is involved in obscurity, and animated by ἐρώs, that is, by Divine Love; from whence proceeded all things which exist."

On LAMARCK (1744-1829) success did not shine as it did on the Comte de Buffon or on Dr. Erasmus Darwin. His life was often so hard that we wonder he did not say more about the struggle for existence. As a youth of sixteen, destined for the Church, he rides off on a bad horse to join the French army, then fighting in Germany, and bravely wins promotion on his first battle-field. After the peace he is sent into garrison at Toulon and Monaco, where his scientific enthusiasm is awakened by the Flora of the south. Retiring in weakened health from military service, he earns his living in a Parisian banker's office, devotes his spare energies to the study of plants, and writes a *Flore française* in three volumes, the publication of which (1778) at the royal press was secured by

U

Buffon's patronage. As tutor to Buffon's son, he travels in Europe and visits some of the famous gardens, and we can hardly doubt that Buffon influenced Lamarck in many ways. After much toil as a literary hack and scientific drudge, he is elected to what we would now call a Professorship of Invertebrate Zoology, a department at that time chaotic. In 1794 he began his lectures, and each year brought increased order to his classification and museum alike. At the same time, however, he was lifting his anchors from the orthodox moorings, relinquishing his belief in the constancy of species, following (we know not with what consciousness) the current which had already borne Buffon and Erasmus Darwin to evolutionary prospects. In 1802 he published *Researches on the Organisation of Living Bodies*; in 1809 a *Philosophie Zoologique*; from 1816-1822 his *Natural History of Invertebrate Animals*, a large work in seven volumes, part of which the blind naturalist dictated to his daughter. Busy as he must have been with zoology, his restless intellect found time to speculate—it must be confessed to little purpose—on chemical, physical, and meteorological subjects. Thus he ran an unsuccessful tilt against Lavoisier's chemistry, and published for ten years annual forecasts of the weather, which seem to have been almost always wrong. Nor did Lamarck add to his reputation by a theory of *Hydrogeology*, and his scientific friends who were loyal specialists shrugged their shoulders more and more over his intellectual knight-errantry.

Poverty also clouded his later years, his treasured collections had to be sold for bread, his theories made no headway, his merits were unrecognised. Yet now a Lamarckian school is strong in France and in America, and even those who deny his doctrines admit that he was one of the bravest of pioneers.

Of Lamarck's *Philosophie Zoologique*, Haeckel says, "This admirable work is the first connected and thoroughly logical exposition of the theory of descent." And again, he says, "To Lamarck will remain the immortal glory of having for the first time established the theory of descent as an independent scientific generalisation of the first order, as the foundation of the whole of Biology." But the verdict of the majority of naturalists in regard to Lamarck's doctrines has not tended to be eulogistic. Cuvier, in his *Éloge de M. de Lamarck* delivered before the French Academy in 1832, said, "A system resting on such foundations may amuse the imagination of a poet, etc., . . . but it cannot for a moment bear the examination of any one who has dissected the hand, the viscera, or even a feather." The great Cuvier was a formidable obscurantist.

But let us hear Lamarck himself:—

"Nature in all her work proceeds gradually, and could not produce all the animals at once. At first she formed only the simplest, and passed from these on to the most complex."

"The limits of so-called species are not so constant and unvarying as is commonly supposed. Spontaneous generation started each particular series, but thereafter one form gives rise to another. In life we should see, as it were, a ramified continuity if certain species had not been lost."

"The operations of Nature in the production of animals show that there is a primary and predominant cause which gives to animal life the power of progressive organisation, of gradually complicating and perfecting not only the organism as a whole, but each system of organs in particular."

"*First Law.* Life by its inherent power tends continually to increase the volume of every living body, and to extend the dimensions of its parts up to a self-regulated limit.

"*Second Law.* The production of a new organ in an animal body results from the occurrence of some new need which continues to make itself felt, and from a new movement which this need originates and sustains.

"*Third Law.* The development of organs and their power of action are constantly determined by the use of these organs.

"*Fourth Law.* All that has been acquired, begun, or changed in the structure of individuals during the course of their life is preserved in reproduction and transmitted to the new individuals which spring from those which have experienced the changes."

These four laws I have cited from Lamarck's *Histoire Naturelle*, but in illustration of the emphasis with which he insisted on use and disuse, I take the following passages, translated by Samuel Butler, from the *Philosophie Zoologique*:—

"Every considerable and sustained change in the surroundings of any animal involves a real change in its needs."

"Such change of needs involves the necessity of changed action in order to satisfy these needs, and, in consequence, of new habits."

"It follows that such and such parts, formerly less used, are now more frequently employed, and in consequence become more highly developed; new parts also become insensibly evolved in the creature by its own efforts from within."

"These gains or losses of organic development, due to use or disuse, are transmitted to offspring, provided they have been common to both sexes, or to the animals from which the offspring have descended."

The historian of the evolution of evolution theories should take account of many workers besides Buffon, Erasmus Darwin, and Lamarck; of Treviranus (1776-1837), whose *Biology or Philosophy of Living Nature* (1802-1805) is full of evolutionary suggestions; of Geoffroy St. Hilaire, who in 1830, before the French Academy of Science, fought with Cuvier, the fellow-worker of his youth, an intellectual duel on the question of descent; of Goethe who, in his eighty-first year, heard the tidings of Geoffroy's defeat with an interest which transcended the political anxieties of the time, and whose own epic of evolution surpasses that of Lucretius; of Oken's speculative mist, amid which the light of evolutionary ideas danced like a will-o'-the-wisp; of many others in whose mind the truth grew if it did not blossom. But we must now recognise the work of Charles Darwin.

6. **Darwin.**—Though the general tenor of Darwin's life—the impression of an industrious open-minded observer and thinker, the picture of a man full of mercy, kindliness, and peace—was familiar to many, his biography has filled in those little details which make our impression living. We see him now, as in a Holbein picture, with all the paraphernalia of daily pursuit round about him. His high chair, his orderly shelves, his torn-up reference books, his window-sill laboratory, his yellow-back novels, his snuff-box, and a hundred little touches, make the picture alive. We learn, too, his methods of laborious but never toilsome work, and the gradual progress of his thought from the conventionalism of youth to the convictions of matured man-

hood. We read the curve of his moods, steadier than that of most men, without any climax of speculative ecstasy, free from any fall to a depth of pessimism. We hear his own sincere voice in his simple autobiography, and even more clearly perhaps in the unconstrainedness of his abundant letters. There was seldom a great life so devoid of littleness, seldom a record of thought so free from subtlety. There seems to be almost nothing hid which we could wish revealed, or uncovered which we could wish hidden. Darwin's life was as open as the country around his hermitage.

Marcus Aurelius gives thanks in his roll of blessings that he had not been suffered to keep quails; so Darwin, in recounting his mercies, does not forget to be grateful for having been preserved from the snare of becoming a specialist. From a more partial point of view, we have reason to be thankful that he became a specialist, not in one department, but in many. As a disciple of Linnæus, he described the species of barnacles in one volume, and followed in the steps of Cuvier in anatomising them in another. Of tissues and cells he knew less, being as regards these items an antediluvian, and outside the guild of those who dexterously wield the razor, and in so doing observe the horoscope of the organism. Of protoplasm, in regard to which modern biology says so much and knows so little, he was not ignorant, for did he not study the marvels of the state known as "aggregation"?

But it is not for special research that men are most grateful to Darwin. Undoubtedly, if clear insight into the world around us be esteemed in itself of value, the author of *Insectivorous Plants*, *The Fertilisation of Orchids*, *The Movements of Plants*, *The Origin of Coral Reefs*, *The Formation of Vegetable Mould*, etc., runs no risk of being forgotten. But though our possession of these results swells the meed of praise, we usually regard them as outside of Darwin's real work, which, as every one knows, was his contribution to the theory of organic life.

This contribution was threefold—(*a*) He placed the theory of descent on a sure basis; (*b*) he shed the light of this doctrine on various groups of phenomena; and (c) he essayed the problem of the factors in evolution.

(a) The man who makes us believe a fact is to us more important than the original discoverer. And so Darwin gets credit for inventing the theory of descent, which in principle is as old as clear thought itself, and in its biological application was stated a hundred years before the publication of the *Origin of Species* (1859). The conception was no new one, but Darwin first made men believe it. The idea was not his, but he gave it to many. He did not originate; he established. He converted naturalists to an evolutionary conception of the organic world.

(b) Having got people to believe the theory of descent,—the theory of development out of preceding conditions,—Darwin went on to show how the conception would illumine all facts to which it was applicable. In his work on the expression of emotions, and in scattered chapters, he showed how the light might be shed upon the secrets of mental activity. Whenever it was seen that the doctrine could justify itself in regard to general organic life, it was eagerly seized as an organon for the exploration of special sets of facts. The phœnix revived and flew croaking amid the smoke of burning systems. How one discussed the evolution of language, and another that of industry; how the natural history of ethics was sketched by one thinker, and the descent of institutions by another; how the conception has forced its way into the cloister and the political arena, and has even found expression in theories of literature, art, and religion,—is an often-repeated story.

(c) We have noticed that Buffon, and, let us add, Treviranus, firmly maintained that the direct influence of the external conditions of life was an important factor in evolution. We have also seen that Erasmus Darwin and Lamarck were strongly convinced of the transforming power of use and disuse. When Charles Darwin began to think and write on the origin of species, he also recognised the transforming influences of function and of environment. But with the Buffonian or Lamarckian position he was never satisfied; he advanced to one of his own—to the theory of natural selection, the characteristic feature of Darwinism.

Let us state this theory, which was foreseen by Matthew, Wells, Naudin, and others, was developed simultaneously by Darwin and by Alfred Russel Wallace, and has attained remarkable acceptance throughout the world.

All plants and animals produce offspring which, though like their parents, usually differ from them in possessing some new features or variations. These are of more or less obscure origin, and are often termed fortuitous or indefinite. But throughout nature there is a struggle for existence in which only a small percentage of the organisms born survive to maturity or reproduction. Those which survive do so because of the individual peculiarities which have made them in some way more fit to survive than their fellows. Moreover the favourable variation possessed by the survivors is handed on as an inheritance to their offspring, and tends to be intensified when the new generation is bred from parents both possessing the happily advantageous character. This natural fostering of advantageous variations and natural elimination of those less fit, explain the general modification and adaptation of species, as well as the general progress from simpler to higher forms of life.

This theory that favourable variations may be fostered and accumulated by natural selection till useful adaptations result is the chief characteristic of Darwinism. Of this theory Prof. Ray Lankester says: " Darwin by his discovery of the mechanical principle of organic evolution, namely, the survival of the fittest in the struggle for existence, completed the doctrine of evolution, and gave it that unity and authority which was necessary in order that it should reform the whole range of philosophy." And again he says: "The history of zoology as a science is therefore the history of the great biological doctrine of the evolution of living things by the natural selection of varieties in the struggle for existence,—since that doctrine is the one medium whereby all the phenomena of life, whether of form or function, are rendered capable of explanation by the laws of physics and chemistry, and so made the subject-matter of a true science or study of causes." I have quoted these two sentences because they illustrate better than any others that I have seen to what exaggeration enthusiasm for a theory will lead a strong intellect. But listen to a few sentences from Samuel Butler, which I quote because they well illustrate that the critics of Darwinism may also be extreme, and in the hope that the contrast may be sufficiently interesting to induce you to think out the question for yourselves.

"Buffon planted, Erasmus Darwin and Lamarck watered, but it was Mr. Darwin who said 'That fruit is ripe,' and shook it into his lap. . . . Darwin was heir to a discredited truth, and left behind him an accredited fallacy. . . . Do animals and plants grow into conformity with their surroundings because they and their fathers and mothers take pains, or because their uncles and aunts go away? . . . The theory that luck is the main means of organic modification is the most absolute denial of God which it is possible for the human mind to conceive. . . ."

7. **Darwin's Fellow-workers.**—But we must bring this historical sketch to a close by referring to four of the more prominent of Darwin's fellow-workers—Wallace, Spencer, Haeckel, and Huxley.

ALFRED RUSSEL WALLACE, contemporary with Darwin, not only in years, but in emphasising the truth of evolutionary conceptions, and in recognising the fact of natural selection, has been justly called the Nestor of Biology. No one will be slow to appreciate the splendid unselfishness with which he has for thirty years sunk himself in the Darwinian theory, or the scientific disinterestedness which leads him from the very title of his last work[1] to its close, to say so little—perhaps too little—of the important part which he has played in evolving the doctrine. "It was," Romanes says, "in the highest degree dramatic that the great idea of natural selection should have occurred independently and in precisely the same form to two working naturalists; that these naturalists should have been countrymen; that they should have agreed to publish their theory

[1] *Darwinism*, London, 1889.

on the same day; and last, but not least, that, through the many years of strife and turmoil which followed, these two English naturalists consistently maintained towards each other such feelings of magnanimous recognition that it is hard to say whether we should most admire the intellectual or the moral qualities which, in relation to their common labours, they have displayed."

Mr. Wallace is a naturalist in the old and truest sense, rich in a world-wide experience of animal life, at once specialist and generaliser, a humanist thinker and a social striver, and a man of science who realises the spiritual aspect of the world.

He believes in the "overwhelming importance of natural selection over all other agencies in the production of new species," differs from Darwin in regard to sexual selection, to which he attaches little importance, and agrees with Weismann in regard to the non-inheritance of acquired characters.

But the exceptional feature in Wallace's scientific philosophy is his contention that the higher characteristics of man are due to a special evolution hardly distinguishable from creation.

Wallace finds their only explanation in the hypothesis of "a spiritual essence or nature, capable of progressive development under favourable conditions."

HERBERT SPENCER must surely have been an evolutionist by birth; there was no hesitation even in the first strides he took with the evolution-torch uplifted. A ponderer on the nature of things, and the possessor of encyclopædic knowledge, he grasped what was good in Lamarck's work, and as early as 1852 published a plea for the theory of organic evolution which is still remarkable in its strength and clearness. The work of Darwin supplied corroboration and fresh material, and in the *Principles of Biology* (1863-66) the theory of organic evolution first found philosophic, as distinguished from merely scientific expression. To Spencer we owe the familiar phrase "the survival of the fittest," and that at first sight puzzling generalisation, "Evolution is an integration of matter and concomitant dissipation of motion, during which the matter passes from an indefinite incoherent homogeneity to a definite coherent heterogeneity, and during which the retained motion (energy) undergoes a parallel transformation." He has given his life to establishing this generalisation, and applying it to physical, biological, psychological, and social facts. As to the factors in organic evolution, he emphasises the change-producing influences of environment and function, and recognises that natural selection has been a very important means of progress.

ERNST HAECKEL, Professor of Zoology in Jena, and author of a great series of monographs on Radiolarians, Sponges, Jellyfish, etc., may be well called the Darwin of Germany. He has devoted his life to applying the doctrine of descent, and to making it current coin among the people. Owing much of his motive to Darwin, he stood for a time almost alone in Germany as the champion of a heresy. Before the publication of Darwin's *Descent of Man*, Haeckel was the only naturalist who had recognised the import of sexual selection; and of his *Natural History of Creation* Darwin writes: "If this work had appeared before my essay had been written, I should probably never have completed it." His most important expository works are the above-mentioned *Natürliche Schöpfungsgeschichte* (1st ed. 1868; 8th ed. 1889); and his *Anthropogenie* (1874, translated as *The Evolution of Man*). These books are very brilliantly written, though they offend many by their remorseless consistency, and by their impatience with theological dogma and teleological interpretation. His greatest work, however, is of a less popular character, namely, the *Generelle Morphologie* (2 vols., Berlin, 1866), which in its reasoned orderliness and clear generalisations ranks beside Spencer's *Principles of Biology*.

HUXLEY, by whose work the credit of British schools of zoology has been for many years enhanced, was one of the first to stand by Darwin, and to wield a sharp intellectual sword in defence and attack. No one has fought for the doctrine of descent in itself and in its consequences with more keenness and success than the author of *Man's Place in Nature* (1863), *American Addresses*, *Lay Sermons*, etc., and no one has championed the theory of natural selection with more confident consistency or with more skilfully handled weapons.

8. **The Present State of Opinion.**—As Wallace says in the preface to his work on *Darwinism*, "Descent with modification is now universally accepted as the order of nature in the organic world." But, while this is true, there remains much uncertainty in regard to the way in which the progressive ascent of life has come about, as to the mechanism of the great nature-loom. The relative importance of the various factors in evolution is very uncertain.[1]

The condition of evolution is variability, or the tendency which animals have to change. The primary factors of evolution are those which produce variations, which cause organic inequilibrium. Darwin spoke of variations as "fortuitous," "indefinite," "spontaneous," etc., and frankly confessed that he could not explain how most of them arose.

Ultimately all variations in organisms must be due to variations in their environment, that is to say, to changes in the system of which organisms form a part. But this is only a general truism.

---

[1] All naturalists, however uncertain in regard to the factors in evolution, accept the doctrine of descent—the general conception of evolution—as a theory which has justified itself. It is not indeed so demonstrable as is the doctrine of the conservation of energy, but it is almost as confidently accepted. Few naturalists, however, have attempted any philosophical justification of their belief. This is strange, since it should surely give pause to the dogmatic evolutionist to reflect that his own theory has been evolved like other beliefs, that his scientific demonstration of it rests upon assumptions which have also been evolved, that the entire system of evolutionary thought must be a phase in the development of opinion, that, in short, he cannot be dogmatic without being self-contradictory. See A. J. Balfour's *Defence of Philosophic Doubt*, pp. 260-274 (London, 1879). In regard to the philosophical aspects of the doctrine of evolution see Prof. Knight's essay on "Ethical Philosophy and Evolution" in his *Studies in Philosophy and Literature* (Lond. 1879), and, with additions, in *Essays in Philosophy* (Boston and New York, 1890); Prof. St. George Mivart's *Contemporary Evolution* (Lond. 1876); E. von Hartmann's *Wahrheit und Irrthum im Darwinismus*; an article by Prof. Tyndall on "Virchow and Evolution" in *Nineteenth Century*, Nov. 1878; and articles on "Evolution" by Huxley and Sully in *Encyclopædia Britannica*.

There are evidently three direct ways in which organic changes may be produced: (1) From the nature of the organism itself; *i.e.* from constitutional or germinal peculiarities which are ultimately traceable to influences from without; (2) from changes in its functions or activity, in other words, from use and disuse; or (3) from the direct influence of the external conditions of life—food, temperature, moisture, etc.

Thus some naturalists follow Buffon in emphasising the moulding influence of the environment, or agree with Lamarck in maintaining that change of function produces change of structure. But at present the tide is against these opinions, because of the widespread scepticism as to the transmissibility of characters thus acquired.

Those who share this scepticism refer the origin of variations to the nature of the organism, to the mingling of the two different cells from which the individual life begins, to the instability involved in the complexity of the protoplasm, to the oscillating balance between vegetative and reproductive processes, and so on.

One prevalent opinion regards variations as arbitrary sports in "a chapter of accidents," but according to the views of a minority variations are for the most part definite, occurring in a few directions, fixed by the constitutional bias of the organism. The minority are "Topsian" evolutionists who believe that the modification of species has taken place by cumulative growth, influenced by function and environment, and pruned by natural selection. To the majority the theory that new species result from the action of natural selection on numerous, spontaneous, indefinite variations, is the "quintessence of Darwinism" and of truth.

Until we know much more about the primary factors which directly cause variations it will not be possible to decide in regard to the precise scope of natural selection and the other secondary factors which foster or accumulate, thin or prune, which in short establish a new organic equilibrium. The argument has been too much in regard to possibilities, too little in regard to observed facts of variation.

The secondary factors of evolution may be ranked under two heads :—

1. Natural Selection, or the survival of the fittest in the struggle for existence, and 2. Isolation, or the various means by which species tend to be separated into sections which do not interbreed.

Natural selection is a phrase descriptive of the course of nature, of the survival of the fit and the elimination of the unfit in the struggle for existence. It involves on the one hand the survival, *i.e.* the nutritive and reproductive success of the variations fittest to survive in given conditions, and on the other hand the destruction or elimination of forms less fit. Suitable variations pay; nature or natural selection justifies and fosters them. Maternal sacrifice or cunning cruelty, the milk of animal kindness or teeth strong to rend, distribution in space or rate of reproduction, are all affected by natural selection. But it is another thing to say that all the adaptations and well-endowed species that we know have been produced by the action of natural selection on fortuitous, indefinite variations. This is what Samuel Butler calls the "accredited fallacy."

Secondly, there seem to be a great many ways by which a species may be divided into two sections which do not interbreed, and if this isolation be common it must help greatly in divergent evolution.

Thus Romanes, who has been the chief exponent of the importance of isolation, on which Gulick has also insisted, says: "Without isolation, or the prevention of free intercrossing, organic evolution is in no case possible. It is isolation that has been 'the exclusive means of modification,' or more correctly, the universal condition to it. Heredity and variability being given, the whole theory of organic evolution becomes a theory of the causes and conditions which lead to isolation."

SUMMARY OF EVOLUTION THEORIES.

	Variations all ultimately due to External Influences.			
Primary Factors	Direct action of the environment produces environmental variations,	Organismal, constitutional, congenital, or germinal variations may be either definite or indefinite.		Use and disuse and change of function produce functional variations
	which if transmissible,	(certainly transmissible).		which, if transmissible,
Secondary Factors	may accumulate as environmental modifications of species.	By the persistence of the original conditions these may grow into new species.	By natural selection in the struggle for existence these may give rise to new species.	may accumulate as functional modifications of species.
	*	*		*
	All cases	may	be affected by	"isolation."

Origin of Variations. Origin of Species.

The process of natural selection will affect all cases, but is less essential for those marked *.

CHAPTER XIX

THE INFLUENCE OF HABITS AND SURROUNDINGS

1. *The Influence of Function*—2. *The Influence of Surroundings*—3. *Our own Environment*

1. **The Influence of Function.**—A skilled observer can often discern a man's occupation from his physiognomy, his shoulders, or his hands. In some unhealthy occupations the death-rate is three times that in others. Disuse of such organs as muscles tends to their degeneration, for the nerves which control them lose their tone and the circulation of blood is affected; while on the other hand increased exercise is within certain limits associated with increased development. *A force de forger on devient forgeron.*

If we knew more about animals we might be able to cite many cases in which change of function produced change of structure, but there are few careful observations bearing on this question.

Even if we could gather many illustrations of the influence of use and disuse on individual animals, we should still have to find out whether the precise characters thus acquired by individuals were transmissible to the offspring, or whether any secondary effects of the acquired characters were transmissible, or whether these changes had no effect upon succeeding generations. As there are few facts to argue from, the answers given to these questions are not reliable.

It is easy to find hundreds of cases in which the constant

characters of animals may be hypothetically interpreted as the result of **use** or disuse. Is the torpedo-like shape of **swift swimmers** due to their rapid motion through the water, do burrowing animals necessarily become worm-like, **has the giraffe** lengthened its neck by stretching it, have **hoofs been** developed by running **on hard** ground, are horns **responses to** butting, are **diverse shapes of** teeth the results of chewing diverse kinds of **food, are cave-animals** blind because they have ceased to use their **eyes, are snails** lop-sided because the shell has fallen to **one** side, **is the** asymmetry in the head of flat fishes due to the efforts made by the ancestral fish to use its lower eye after it had formed **the** habit of lying flat **on the bottom, is the** woodpecker's long tongue the **result** of continuous probing into holes, are webbed feet due **to** swimming efforts, has the food-canal in vegetarian animals **been** mechanically lengthened, do the wing bones and muscles **of** the domesticated duck compare unfavourably with those of the wild duck because the habit of sustained flight has been lost by the former?

But these interpretations have not been verified; they are only probable. "It is infinitely easy," Semper says, "to form a fanciful idea as to how this or that fact may be hypothetically explained, and very little trouble is needed to imagine some process by which hypothetical fundamental causes—equally fanciful—may have led to the result which has been actually observed. But when we try to prove by experiment that this imaginary process of development is indeed the true and inevitable one, much time and laborious **research** are indispensable, **or we find** ourselves wrecked on insurmountable difficulties."

Not a few naturalists believe in the inherited effects **of functional** change mainly because the theory is simple and logically sufficient. If use and disuse alter the structure of individuals, if the results are transmitted and accumulate in similar conditions for generations, we require no other explanation of many structures.

The reasons why not a few naturalists disbelieve in the inherited effects of functional change are (1) that definite proof is wanting, (2) that it is difficult to understand how

CH. XIX *Influence of Habits and Surroundings* 305

changes produced in the body by use or disuse can be transmitted to the offspring, (3) that the theory of the accumulation of (unexplained) favourable variations in the course of natural selection seems logically sufficient. I should suspend judgment, because it is unprofitable to argue when ascertained facts are few.

But if you like to argue about probabilities, the following considerations may be suggestive :—

The natural powers of animals—horses, dogs, birds, and others—can be improved by training and education, and animals can be taught tricks more or less new to them, but we have no precise information as to any changes of structure associated with these acquirements.

Individual animals are sometimes demonstrably affected by use or disuse. Thus Packard cites a few cases in which some animals—usually with normal eyes—have had these affected by disuse and darkness; he instances the variations in the eyes of a Myriapod and an Insect living in partial daylight near the entrance of caves, the change in the eyes of the common Crustacean *Gammarus pulex* after confinement in darkness, the fact that the eyes of some other Crustaceans in a lake were smaller the deeper the habitat.

There are many more or less blind animals, and Packard says "no animal or series of generations of animals, wholly or in part, can lose the organs of vision unless there is some appreciable physical cause for it." If so, it is probable that the appreciable physical cause has been a *direct* factor in producing the blindness.

Not a few young animals have structures, such as eyes and legs, which are not used and soon disappear in adult life. Thus the little crab *Pinnotheres*, which lives inside bivalves and sea-cucumbers, keeps its eyes until it has established itself within its host. Then they are completely covered over and degenerate. The same is true of many internal parasites, and Semper concludes that "we must refer the loss of sight to disuse of the organ." Perhaps the same is true of some blind cave-animals, in which the eyes are less degenerate in the young, and of the mole, whose embryos have between the eyes and the brain normal optic nerves which usually degenerate in each individual lifetime.

The theory that many structures in animals are due to the inherited results of use and disuse has this advantage, that it suggests a primary cause of change, whereas the other theory assumes the occurrence of favourable variations and proceeds to show how they might be accumulated in the course of natural selection, that is to say by a secondary factor in evolution.

When we find in a large number of entirely distinct forms that the same habit of life is associated with the same peculiarities, there is a likelihood that the habit is a direct factor in evolving these. Thus sluggish and sedentary animals in many different classes tend to develop skeletons of lime, as in sponges, corals, sedentary worms, lamp-shells, Echinoderms, barnacles, molluscs. Professor Lang has recently made a careful study of sedentary creatures, and this result at least is certain that the same peculiarity often occurs in many different types with little in common except that they are sedentary. But till one can show that sedentary life necessarily involves for instance a skeleton of lime or something equivalent, we are still dealing only with probabilities.

2. **The Influence of Surroundings.**—In ancient times men saw the threads of their life passing through the hands of three sister-fates—of one who held the distaff, of another who offered flowers, and of a third who bore the abhorred shears of death. In Norseland the young child was visited by three sister Norns, who brought characteristic gifts of past, present, and future, which ruled the life as surely as did the hands of the three Fates. So too in days of scientific illumination, we think of the dread three, but, clothing our thoughts in other words, speak of life as determined by the organism's legacy or inheritance, by force of habit or function, and by the influences of external conditions or environment. What the living organism is to begin with, what it does or does not in the course of its life, and what surrounding influences play upon it,—these are the three Fates, the three Norns, the three Factors of Life. Organism, function, and environment are the sides of the bio-

logical prism. Thus we try to analyse the light of life. But inheritance in its widest sense is only another name for the organism itself, and function is simply the organism's activity. The organism is real; the environment is real, in it we live and move; function consists of action and reaction between these two realities. Yet the capital which the organism has to begin with is very important; conduct has some relation to character, and function to structure; the surroundings—the dew of earth and the sunshine of heaven —silently mould the individual destiny.

A living animal is almost always either acting upon its surroundings or being acted upon by them, and life is the relation between two variables—a changeful organism and a changeful environment. And since animals do not and cannot live *in vacuo*, they should be thought of in relation to their surroundings. You may kill the body and cut it to pieces, and the result may be interesting, but you have lost the animal just as you lose a picture if you separate figure from figure, and all from the associated landscape or interior. The three Fates are sisters, they are thoroughly intelligible only as a Trinity.

The most certain of all the relations between an organism and its surroundings is the most difficult to express. We see a small whirlpool on a river, remaining for days or weeks apparently constant, with the water circling round unceasingly, bearing the same flotsam of leaves and twigs. But though the eddy seems the same for many days, it is always changing, currents are flowing in and out; it is the constancy of the stream and its bed which produces the apparent constancy of the whirlpool. So, in some measure, is it with an animal in relation to its surroundings. Streams of matter and energy are continually passing in and out. Though we cannot see it with our eyes, the organism is indeed a whirlpool. It is ever being unmade and remade, and owes much of its apparent constancy to the fact that the conditions in which it lives—the currents of its stream —are within certain limits uniform.

But as we cannot understand the material aspects of an animal's life without considering the streams of matter and energy which pass in and out, neither can we understand its higher life apart from its surroundings.

To attempt a natural history of isolated animals, whether alive or dead, is like trying to study man apart from society. For it is only when we know animals as they live and move that we discover how clever, beautiful, and human they are. Thus Gilbert White's *Selborne* is a *natural* history; and therefore we began our studies with the natural life of animals—their competition and helpfulness, their adaptations to diverse kinds of haunts, their shifts and tricks, their industries and their loves.

At present, however, we have to do with the relation between external and internal changes. We must find out what the environment of an organism is, and what power it has. In a smithy we see a bar of hot iron being hammered into useful form. Around a great anvil are four smiths with their hammers. Each smites in his own fashion as the bar passes under his grasp. The first hammer falls, and while the bar is still quivering like a living thing it receives another blow. This is repeated many times till the thing of use is perfected. By force of smiting one becomes a smith, and by dint of blows the bar of iron becomes an anchor. So is it with the organism. In its youth especially, it comes under the influence of nature's hammers; it may become fitter for life, or it may be battered out of existence altogether. Let us try to analyse the various environmental factors.

(*a*) *Pressures.*—First we may consider those lateral and vertical pressures due to air or water currents and to the gentle but potent force of gravity. The shriek of the wind as it prunes the trees, the swish of the water as it moulds the sponges and water-leaves, illustrate the tunes of those pressure-hammers. Under artificial pressure embryos have been known to broaden; even the division of the egg is affected by gravity; water currents mould shells and corals. The influence of want of room must also be noticed, for by artificial overcrowding naturalists have slowed the rate of development and reared dwarf broods; and the rate of human mortality sometimes varies with the size of the dwelling. It is difficult, however, to abstract the influence of restricted space from associated abnormal conditions.

(b) *Chemical Influences.*—Quieter, but more potent, are the chemical influences which damp or fan the fire of life, which corrode the skin or drug the system, which fatten or starve, depress or stimulate. Along with these we must include that most important factor—food.

When a lighted piece of tinder is placed in a vessel full of oxygen it burns more actively. Similarly, superabundance of oxygen makes insects jump, makes the simplest animals more agile, and causes the "phosphorescent" lights of luminous insects to glow more brightly; and young creatures usually develop more or less rapidly according as the aëration is abundant or deficient. The most active animals—birds and insects—live in the air and have much air in their bodies; sluggish animals often live where oxygen is scarce; changes in the quality of the atmosphere may have been of importance in the historical evolution of animals. Fresh air influences the pitch of human life, and lung diseases increase in direct ratio to the amount of crowded indoor labour in an area.

By keeping tadpoles in unnatural conditions the usual duration of the gilled stage may be prolonged for two or three years. The well-known story of the Axolotl and the *Amblystoma* is suggestive but not convincing of the influence of surroundings. These two newt-like Amphibians differ slightly from one another, in this especially that the Axolotl retains its gills after it has developed lungs, while the *Amblystoma* loses them. Both forms may reproduce, and they were originally referred to different genera. But some Axolotls which had been kept with scant water in the *Jardin des Plantes* in Paris turned into the *Amblystoma* form; the two forms are different phases of the same animal. It was a natural inference that the Axolotls were those which had remained or had been kept in the water, the *Amblystoma* forms were those which got ashore. But both kinds may be found in the water of the same lake and the metamorphosis may take place in the water as well as on the shore. For these and for other reasons this oft-told tale is not cogent. In another part of this book I have given examples of the state of lifelessness which drought induces in some

FIG. 64.—Axolotl (in the water) and Amblystoma (on the land).

simple animals, and from which returning moisture can after many days recall them.

Changes may also be due to the chemical composition of the medium, as was established by the experiments of

FIG. 65.—Side view of male *Artemia salina* (enlarged).
(From Chambers's *Encyclop.*)

Schmankewitsch on certain small Crustaceans. Among the numerous species of the brine-shrimp *Artemia*, the most unlike are *A. salina* and *A. milhausenii*; they differ in the

CH. XIX Influence of Habits and Surroundings 311

shape and size of the tail and in the respiratory appendages borne by the legs; they are not found together, but live in pools of different degrees of saltness. Now Schmankewitsch took specimens of *A. salina* which live in the less salt water,

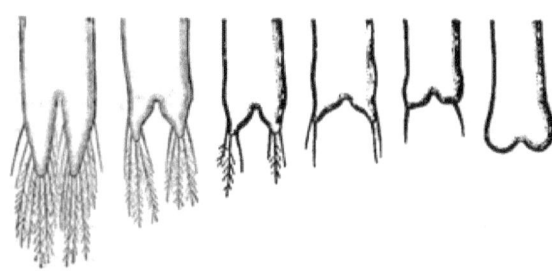

FIG. 66.—Tail-lobes of *Artemia salina* (to the left) and of *Artemia milhausenii* (to the right); between these four stages in the transformation of the one into the other. (From Chambers's *Encyclop.*; after Schmankewitsch.)

added salt gradually to the medium in which they were living, and in the course of generations turned them into *A. milhausenii*. He also reversed the process by freshening the water little by little. Moreover, he accustomed *A. salina* to entirely fresh water, and then found that the form had changed towards that of a related genus, *Branchipus*. This last step has been adversely criticised, but it is allowed that one species of brine-shrimp was changed into another.

Many interesting experiments have been made on the effect of chemical reagents on cells, but these are perhaps of most interest to the student of drugs. Still the fact that the form of a cell and its predominant phase of activity may be entirely changed in this way is important, especially when we remember that it was in single cells that life first began, and is now continued. Even Weismann agrees with Spencer's conclusion that "the direct action of the medium was the *primordial* factor of organic evolution."

To Claude Bernard, the main problem of evolution seemed to be concerned with variations in nutrition: "L'évolution, c'est l'ensemble constant de ces alternatives de la nutrition; c'est la nutrition considerée dans sa réalité, embrassée d'un coup d'œil à travers le temps." John Hunter and others have shown how the walls of the stomach of gulls and other birds may be experimentally altered by change of diet, and the same is seen in nature when the Shetland gull changes from its summer diet of grain to its winter diet of fish. The colours of birds' feathers, as in canaries and parrots, are affected by their food. A slight difference in the quantity and quality of food determines whether a bee-grub is to become a queen or a worker, royal diet evolving the reproductive queen, sparser less rich diet evolving the more active but unfertile worker. Abundant food favours the production of female offspring, while sparser food tends to develop males. Thus, in frogs, the proportion of the sexes is normally not very far from equal; in three lots of tadpoles an average of 57 per hundred became females, 43 males. But Yung has shown that the nutrition of the tadpoles has a remarkable influence on the sex of the adults. In a set of which one half kept in natural conditions developed into 54 females to 46 males, the other half fed with beef had 78 females to 22 males. In a second set of which one half left to themselves developed 61 females to 39 males, the other half, fed with fish, had 81 females to 19 males. Finally, in a third set, of which one half in natural conditions developed 56 females to 44 males, the other half, to which the especially nutritious flesh of frogs was supplied, had no less than 92 females to 8 males.

When food is abundant, assimilation active, and income above expenditure, the animal grows, and at the limit of growth in lower animals asexual multiplication occurs. Checked nutrition, on the other hand, favours the higher or sexual mode of multiplication. Thus the gardener prunes the roots of a plant to get better flowers or reproductive leaves. The plant-lice or Aphides, which infest our pear-trees and rose-bushes, well illustrate the combined influence of food and warmth. All through the summer, when food is abundant and the warmth pleasant, the Aphides enjoy prosperity, and multiply rapidly. For an Aphis may bring forth young every few hours for days together, so rapidly that if all the offspring of a mother Aphis survived, and multiplied as she did, there would in the course of a year be a progeny which would weigh down 500,000,000 stout men. But all through the summer these Aphides are wholly female, and therefore wholly parthenogenetic; no males occur. In autumn, however, when hard times set in, when food is scarcer, and the weather colder, males are born, parthenogenesis ceases, ordinary sexual reproduction recurs. Moreover, if the Aphides be kept in the artificial summer of a greenhouse, as has been done for four years, the parthenogenesis continues without break, no males being born to enjoy the comforts of that environment. Periods of fasting occur in the life-history of many animals, and these are very momentous and progressive periods in the lives of some, for the tadpole fasts before it becomes a frog, and the chrysalis before it becomes a butterfly. Lack of food, however, may stunt development, as we see every day in the streets of our towns.

(c) *Radiant Energy.*—Of the forms of radiant energy which play upon the organism, we need take account only of heat and light, for of electrical and magnetic influence the few strange facts that we know do not make us much wiser.

We know that increased warmth hastens motion, the development of embryos, and the advent of sexual maturity. An Infusorian (*Stylonichia*) studied by Maupas was seen to divide once a day at a temperature of $7°$-$10°$ C., twice at $10°$-$15°$, thrice at $15°$-$20°$, four times at $20°$-$24°$, five times at $24°$-$27°$ C. At the last temperature one Infusorian became in four days the ancestor of a million, in six days of a billion, in seven days and a half of 100 billions, weighing 100 kilogrammes. By consummately patient experiments, Dallinger was able to educate Monads which lived normally at a temperature of $65°$ Fahr., until they could flourish at $158°$ Fahr.

Cold has generally a reverse action, checking activity, producing coma and lifelessness, diminishing the rate of development, tending to produce dwarf or larva-like forms. The cold of winter acting through the nervous system changes the colour of some animals, like Ross's lemming, to advantageous white. Not a few animals vary slightly with the changing seasons. Thus many cases are known where a butterfly produces in a year more than one brood,

FIG. 67.—Seasonal dimorphism of *Papilio ajax*; to the left the winter form (variety *Telamonides*), to the right the summer form (variety *Marcellus*). (From Chambers's *Encyclop.*; after Weismann.)

of which the winter forms are so different from those born in summer that they have often been described as different species. It is possible that this is a reminiscence of past climatic changes, such as those of the Ice Ages, as the

FIG. 68.—Seasonal changes of the bill in the puffin (*Fratercula arctica*); to the left the spring form, to the right the winter form, both adult males. (After Bureau.)

result of which a species became split up into two varieties. Thus *Araschnia levana* and *Araschnia prorsa* are respectively the winter and summer forms of one species. In the glacial epoch there was perhaps only *A. levana*, the winter form; the change of climate has perhaps evolved the summer variety *A. prorsa*. Both Weismann and Edwards have succeeded, by artificial cold, in making the pupæ which should become the summer *A. prorsa* develop into the winter *A. levana*. Nor can we forget the seasonal moulting and the subsequent change of the plumage in birds, so marked in the case of the ptarmigan, which moults three times in the year. In the puffins even the bill is moulted and appears very different at different seasons. But in these last cases the influence of environment must be very indirect.

Light is very healthful, but it is not easy to explain its precise influence. Our pulses beat faster when we go out into the sunlight. Plants live in part on the radiant energy of the sun, and perhaps some pigmented animals do the same. Perhaps the hundreds of eyes which some molluscs have are also useful in absorbing the light. It is also possible that light has a direct influence on the formation of some animal pigments, as it seems to have in the development of chlorophyll. We know, from Poulton's experiments, that the light reflected from coloured bodies influences the colouring of caterpillars and pupæ, but this influence seems to be subtle and indirect, operating through the nervous system. It is also certain that living in darkness tends to bleach some animals, and it is probable that the absence of light stimulus has a directly injurious effect upon the eyes of those animals which live in caves or other dark places. But I have already explained why dogmatism in regard to these cases should be avoided.

One case of the influence of light seems very instructive. It is well known that flat fishes like flounders, plaice, and soles lie or swim in adult life on one side. This lower side is unpigmented; the upper side bears black and yellow pigment-containing cells.

One theory of the presence of pigment on the upper side and its absence on the other is that the difference is a protective adaptation evolved by the natural selection of indefinite variations. But it is open to question whether the characteristic is so advantageously protective as is usually imagined: thus the coloured upper side in soles is very often covered with a layer of sand. Soles come out most at night, most live at depths at which differences of colour are probably indistinct. In shallower water the advantage is likely to be greater, though the white under-side slightly exposed as the fish rises from the bottom may attract attention disadvantageously. Moreover, if we find in a large number of different animals that the side away from the light is lighter than that which is exposed, and if we can show that this has in many cases no protective advantage whatever—and I believe that a few hours' observation will convince you that both my assumptions are correct—then there is a probability that the absence of light has a direct influence on the absence of pigment.

But we are not left to vague probabilities; Mr. J. T. Cunningham has recently made the crucial experiment of illuminating the under sides of young flounders. Out of thirteen, whose under-sides were thus illumined by a mirror for about four months, only three failed to develop black and yellow colour-cells on the skin of the under-sides. It is therefore likely that the normal whiteness of the under-sides is due in some way to the fact that in nature little light can fall on them, for they are generally in contact with the ground.

(*d*) *Animate Surroundings.*—We have given a few instances showing how mechanical or molar pressures, chemical and nutritive influences, and the subtler physical energies of heat and light, affect organisms. There is a fourth set of environmental factors—the direct influence of organism upon organism. In a previous chapter we spoke of the indirect influences different kinds of organisms exert on one another, and these are most important, but there are also results of direct contact.

Much in the same way as insects produce galls on plants, so sea-spiders (*Pycnogonidæ*) affect hydroids, a polype deforms a sponge, a little worm (*Myzostoma*) makes galls on Crinoids. Prof. Giard has described how certain degenerate Crustaceans parasitic on crabs injuriously affect their hosts, and some internal parasites produce slight modifications of structure. Interesting also are the shelters or domatia of some plants, within which insects and mites find homes.

We can speak more confidently about the influence of surroundings than we could in regard to the influence of use and disuse, because the ascertained facts are more numerous. Those interested in the theoretical importance of these facts should attend to the following considerations.

It is essential to distinguish between cases in which we know that external conditions influence the organism and those in which we think they may have done so. Thus it is probable that the degeneracy and other peculiarities of many parasites are results of external influence and of feeding, and also in part of disuse, but we cannot state this as a fact.

Most of the observations on the influence of external conditions give us no information as to the transmissibility of the results. It is not enough to know that a peculiarity *observed to occur* in peculiar surroundings was *observed to recur* in successive generations living in the same surroundings. For (1) it might be an indefinite variation—a sport due to some germinal peculiarity—which happened to suit. In such a case it would be transmissible, but it would not be a change due to the environment. And (2) even when it has been proved that the peculiarity is due to the direct influence of the environment, and observed to recur in successive generations, still its transmissibility is not proven, for it may be hammered on each successive generation as it was on the first. We can say little about the transmissibility or evolutionary importance of changes of structure due to surroundings because most of the observations were made before the scepticism as to the inheritance of acquired characters became dominant. Only in a few cases, such as that of the brine-shrimps, was the cumulative influence traced through many generations. In dearth of facts we should not be confident, but eager for experiment.

Surroundings may influence the organism in varying

degrees. There may be direct results, rapid parries after thrusts, or the results may be indirect; they may affect the organism visibly in the course of one generation, or only after several have passed.

Some animals are more susceptible and more plastic than others. Young organisms, such as caterpillars and tadpoles, are more completely in the grasp of their environment than are the adults. Thus Treviranus, who believed very strongly in the influence of surroundings, distinguished two periods of *vita minima*—in youth and in old age—during which external conditions press heavily, from the period of *vita maxima*—in adult life—when the organism is more free. To some kinds of influence, *e.g.* mechanical pressures, passive and sedentary organisms such as sponges, corals, shell-fish, and plants, are more susceptible than are those of active life. And it is during a period of quiescence that surrounding colour tells on the sensitive caterpillars.

3. **Our own Environment.**—The human organism, like any other, may be modified by its environment, for we lead no charmed life. Those external influences which touch body and mind are to us the more important, since we have them to some extent within our own hands, and because our lives are relatively long. Even if the changes thus wrought upon parents are not transmissible, it is to some extent possible for us to secure that our children grow up open to influences known to be beneficial, sheltered from forces known to be injurious.

As the influence of surroundings is especially potent on young things—such as caterpillars and tadpoles—all care should be taken of the young child's environment during the earliest months and years, when the grip that externals have is probably much greater than is imagined by those who believe themselves emancipated from the tyranny of the present.[1]

As passive organisms are more in the thrall of their surroundings than are the more active, we feel the importance of beauty in the home, that the organism may be

---

[1] Cf. Matthew Arnold's poem, "The Future," and Walt Whitman's "Assimilations."

saturated with healthful influence during the periods in which it is most susceptible. The efforts of Social Unions, Kyrle Societies, Verschönerungs-Vereine, and the like, are justified not only by their results,[1] but by the biological facts on which they more or less unconsciously depend. There would be more progress and less invidious comparison of ameliorative schemes, if we realised more vividly that the Fates are three. Though it is not easy to appreciate the three sides of a prism at once, of what value is liberty on an ash-heap, or equality in a hell, or fraternity among an overpopulated community of weaklings? Organism, function, and environment must evolve together, and surely they shall.

Poets have often compared human beings to caterpillars; it may be that no improvement in constitutions, functions, or surroundings will make us winged Psyches, yet it may be possible for us to be ennobled like those creatures which in gilded surroundings became golden. Surely art is warranted by the results of science, as these in time may justify themselves in art.

[1] Ideally stated in Emerson's well-known poem of "Art."

CHAPTER XX

HEREDITY

1. *The Facts of Heredity*— 2. *Theories of Heredity: theological, metaphysical, mystical, and the hypothesis of pangenesis*— 3. *The Modern Theory of Heredity*— 4. *The Inheritance of Acquired Characters*—5. *Social and Ethical Aspects*—6. *Social Inheritance*

WE have spoken of the three Fates which were believed to determine of what sort a life should be. With the decay of poetic feeling, and in the light of common science, the forms of the three sisters have faded. But they are realities still, for men are thinking more and more vividly about the factors of life, which to some are "powerful principles," to others living and personal, to others unnameable. Biologists speak of them as Heredity, Function, and Environment: the capital with which a life begins, the interest accruing from the investment of this in varied vital activities, and the force of circumstances. But while it is useful to think of Heredity, Function, and Environment as the three fates, we must not mystify matters by talking as if these were entities acting upon the organism. They are simply aspects of the fact that the animal is born and lives. The inheritance is the organism itself, and heredity is only a name for the relation between successive generations. Moreover, the function of an organism depends upon the nature of the organism, and so does its susceptibility to influences from without.

I would at present define heredity as *the organic relation between successive generations*, choosing this definition because it is misleading to talk about "heredity" as a "basal principle in evolution," as a "great law," as a "power," or as a "cause." When I call heredity a "Fate," it is plain that I speak fancifully, but "principle" and "law" are dangerous words to play with. We cannot think of life without this organic relation between parents and offspring, and had species been created instead of being evolved there would still be heredity.

1. **The Facts of Heredity.** — An animal sometimes arises as a bud from its parent, and in rare cases from an egg which requires no fertilisation, but apart from these exceptions, every animal develops from an egg-cell with which a male-cell has united in an intimate way. The egg-cell supplies most of the living matter, but the nucleus of the fertilised egg-cell is formed in half from the nucleus of the immature ovum, in half from the nucleus of the spermatozoon. Let us emphasise this first fact that each parent contributes the same amount of nuclear material to the offspring, and that this nuclear stuff is very essential.

Another fact is more obvious, the offspring is very like its kind. One of the first things that people say about an infant is that it is like its father or its mother, and the assertion does not arouse any surprise, although the truer verdict that the infant is like any other of the same race is received with contempt. But every one admits that "like begets like."

This likeness between offspring and parent is often far more than a general resemblance, for peculiar features and minute idiosyncrasies are frequently reproduced. Yet one must not assume that because a child twirls his thumbs in the same way as his father did the habit has been inherited. For peculiar habits and structures may readily reappear by imitation, or because the offspring grow up in conditions similar to those in which the parents lived.

Abnormal as well as normal characters, "natural" to the parents, may reappear in their descendants, and the list of weaknesses and malformations which may be transmitted is long and grim. But care is required to distinguish

Y

between reappearance due to inheritance and reappearance due to similar conditions of life.

Then there is a strange series of facts showing that an organism may reproduce characteristics which the parents did not exhibit, but which were possessed by a grandparent or remoter ancestor. Thus a lizard in growing a new tail to replace one that has been lost has been known to grow one with scales like those of an ancestral species. To find out a lizard's pedigree, a wit suggests that we need only pull off its tail. When such ancestral resemblance in ordi-

FIG. 69.—Devonshire pony, showing the occasional occurrence of ancestral stripes. (From Darwin.)

nary generation is very marked, we call it "atavism" or "reversion," but of this there are many degrees, and abnormal circumstances sometimes force reversion even upon an organism with a normal inheritance. A boy "takes after his grandfather"; a horse occasionally exhibits stripes like those of a wild ancestor; a blue pigeon like the primitive rock-dove sometimes turns up unexpectedly in a pure breed; or a cultivated flower reverts to the simpler and more normal wild type. So children born during famine sometimes show reversions, and some types of criminal and insane persons are to be thus regarded.

But every animal is usually a little different from its parents, and except in cases of "identical twins" cannot be mistaken for one of its fellow-offspring. The proverbial "two peas" may be very unlike. Organisms are variable, and this is natural, for life begins in the intimate mingling of two units of living matter perhaps very different and certainly very complex. The relation between successive generations is such that the offspring is like its parents, but various causes producing change diminish this likeness, so that we no longer say "like begets like," but "like *tends* to beget like."

There are, I think, two other important facts in regard to heredity, but both require discussion—the one because some of the most authoritative naturalists deny it, the other because it is difficult to understand.

I believe that some characters acquired by the parent as the result of what it does, and as impacts from the surrounding conditions of life, are transmissible to the offspring. In other words, some functional and environmental variations in the body of the parents may be handed on to the offspring. This is denied by Weismann and many others.

The other fact, which has been elucidated by Galton, is that through successive generations there is a tendency to sustain the average of the species, by the continual approximation of exceptional forms towards a mean.

2. **Theories of Heredity—historical retrospect.—**Theories of heredity, like those about many other facts, have been formulated at different times in different kinds of intellectual language—theological, metaphysical, and scientific—and the words are often more at variance than the ideas.

(*a*) *Theological Theories.*—It was an old idea, that the germ of a new human life was possessed by a spirit, sometimes of second-hand origin, having previously belonged to some ancestor or animal. So far as this idea persists in the minds of civilised men, it is so much purified and sublimed that if the student of science does not believe it true, he cannot wisely call it false.

(*b*) "*Metaphysical Theories.*"—For a time it was common to appeal to "*vires formativæ*," "hereditary tendencies," and "principles of heredity," by aid of which the germ grew into the likeness of the parent, and this tendency to resort to verbal explanations is hardly to be driven from the scientific mind except by intellectual asceticism. For my own part, I prefer such "metaphysical" mist to the frost of a "materialism" which blasts the buds of wonder.

(*c*) "*Mystical Theories.*"—During the eighteenth century and even within the limits of the enlightened nineteenth, a quaint idea of development prevailed, according to which the germ (either the ovum or the sperm) contained a miniature organism, preformed in all transparency, which only required to be unfolded (or "evolved," as they said), in order to become the future animal. Moreover, the egg of a fowl contained not only a micro-organism or miniature model of the chick, but likewise in increasing minuteness similar models of future generations. Microcosm lay within microcosm, germ within germ, like the leaves within a bud awaiting successive unfolding, or like an infinite juggler's box to the "evolution" of which there was no end. This "preformation theory" or "mystical hypothesis" was virtually but not actually shattered by Wolff's demonstration of "Epigenesis" or gradual development from an apparently simple rudiment. But the preformationists were right in insisting that the future organism lay (potentially) within the germ, and right also in supposing that the germ involved not only the organism into which it grew but its descendants as well. The form of their theory, however, was crude and false.

(*d*) *Theories of Pangenesis.*—Scientific theories of heredity really begin with that of Herbert Spencer, who in 1864 suggested that "physiological units" derived from and capable of growth into cells were accumulated from the body into the reproductive elements, there to develop the characters of structures like those whence they arose. At dates so widely separate as are suggested by the names of Democritus and Hippocrates, Paracelsus and Buffon, the same idea was expressed—that the germs consist of samples from the various parts of the body. But the theories of these authors were vague and in some respects entirely erroneous suggestions. The best-known form of this type of theory is Darwin's "provisional hypothesis of pangenesis" (1868), according to which (*a*) every cell of the body, not too highly differentiated, throws off characteristic gemmules, which (*b*) multiply by fission, retaining their peculiarities, and (*c*) become specially concentrated in the reproductive elements, where (*d*) in development they grow into cells like those from which they were originally given off. This theory was satisfactory in giving a reasonable explanation of many of the facts of heredity, it was unsatisfactory because it involved many unverified hypotheses.

The ingenious Jæger, well known as the introducer of comfortable clothing, sought (1876) to replace the "gemmules" of which Darwin spoke, by characteristic "scent-stuffs," which he supposed to be collected from the body into the reproductive elements.

Meanwhile (1872) Francis Galton, our greatest British authority on heredity, had been led by his experiments on the transfusion of blood and by other considerations to the conclusion that "the doctrine of pangenesis, pure and simple, is incorrect." As we shall see, he reached forward to a more satisfactory doctrine, but he still allowed the possibility of a limited pangenesis to account for those cases which suggest that some characters acquired by the parents are "faintly heritable." He admitted that a cell "may throw off a few germs" (*i.e.* "gemmules") "that find their way into the circulation, and have thereby a chance of occasionally finding their way to the sexual elements, and of becoming naturalised among them."

W. K. Brooks, a well-known American naturalist, proposed in 1883 an important modification of Darwin's theory, especially insisting on the following three suppositions: that it is in *unwonted* and *abnormal* conditions that the cells of the body throw off gemmules; that the *male* elements are the special centres of their accumulation; and that the *female* cells keep up the general resemblance between offspring and parents. For further modifications and for criticism of the theories of pangenesis, I refer the student to the works of Galton, Ribot, Brooks, Herdman, Plarre, Van Bemmelen, and De Vries.

3. **The Modern Theory of Heredity.**—In the midst of much debate it may seem strange to speak of the modern theory of heredity, but while details are disputed, one clear fact is generally acknowledged, the increasing realisation of which has shed a new light on heredity. This fact is the organic continuity of generations.

In 1876 Jæger expressed his views explicitly as follows: "Through a long series of generations the germinal protoplasm retains its specific properties, dividing in development into a portion out of which the individual is built up, and a portion which is reserved to form the reproductive material of the mature offspring." This reservation, by which some of the germinal protoplasm is kept apart, during development and growth, from corporeal or external influences, and retains its specific or germinal characters intact and continuous with those of the parent ovum, Jæger regarded as the fundamental fact of heredity.

Brooks (1876, 1877, 1883) was not less clear: "The ovum gives rise to the divergent cells of the organism, but also to cells like itself. The ovarian ova of the offspring are these latter cells or their direct unmodified descendants. The ovarian ova of the offspring thus share by direct inheritance all the properties of the fertilised ova."

But before and independently of either Jæger or Brooks or any one else, Galton had reached forward to the same idea. We have noticed that he was led in 1872 to the conclusion that "the doctrine of pangenesis, pure and simple, is incorrect." His own view was that the fertilised ovum consisted of a sum of germs, gemmules, or organic units of some kind, to which in entirety he applied the term *stirp*. But he did not regard this nest of organic units as composed of contributions from all parts of the body. He regarded it as directly derived from a previous nest, namely, from the ovum which gave rise to the parent. He maintained that in development the bulk of the stirp grew into the body—as every one allows—but that a certain residue was kept apart from the development of the "body" to form the reproductive elements of the offspring. Thus he said, in a sense the child is as old as the parent, for when the parent is developing from the ovum a residue of that ovum is kept apart to form the germ-cells, one of which may become a child. Besides Galton, Jæger, and Brooks, several other biologists suggested this fertile idea of the organic continuity of generations. Thus it is expressed by Erasmus Darwin and by Owen, by Haeckel, Rauber, and Nussbaum. But it is to Weismann that the modern emphasis on the idea is chiefly due.

Let us try to realise more vividly this doctrine of organic continuity between generations. Let us begin with a fertilised egg-cell, and suppose it to have qualities $abcxyz$. This endowed egg-cell divides and redivides, and for a short time each of the units in the ball of cells may be regarded as still possessed of the original qualities $abcxyz$. But division of labour, and rearrangement, infolding and outfolding, soon begin, and most of the cells form the "body." They lose their primitive characters and uniformity, they become specialised, the qualities ab predominate in one set, bc in another, xy in another. But meantime certain cells have kept apart from the specialisation which results in the body. They have remained embryonic and undifferentiated, retaining the many-sidedness of the original egg-cell, preserving intact the qualities $abcxyz$. They form the future reproductive cells—let us say the eggs.

Now when these eggs are liberated, with the original qualities $abcxyz$ unchanged, having retained a continuous protoplasmic tradition with the parent ovum, they are evidently in almost the same position as that was. Therefore they develop into the same kind of organism. Given the same protoplasmic material, the same inherent qualities, the same conditions of birth and growth, the results *must* be the same. A single-celled animal with qualities $abcxyz$ divides into two; each has presumably the qualities of the original unit; each grows rapidly into the form of the full-grown cell. We have no difficulty in understanding this. In the sexual reproduction of higher animals, the case is complicated by the formation of the "body," but logically the difficulty is not greater. A fertilised egg-cell with qualities *abcxyz* divides into many cells, which, becoming diverse, express the original qualities in various kinds of tissue within the forming body. But if at an early stage certain cells are set apart, retaining the qualities or characters *abcxyz* in all their entirety, then these, when liberated after months or years as egg-cells, will resemble the original ovum, and are able like it to give rise to an organism, which is necessarily a similar organism.

To call heredity "the relation of organic continuity between successive generations," as I define it, seems a truism to some, but it is in the realisation of this truistic fact that the modern progress in regard to heredity consists.

To ask how the inherent qualities of the ovum become divergent in the different cells of the body, or how some units remain embryonic, or how the egg-cell divides at all, is to raise the deepest problems of biology, not of heredity. To answer such questions is the more or less hopeless task of physiological embryology, not that of the student of heredity. Recognising the fact of organic continuity, various writers such as Samuel Butler, Hering, Haeckel, Geddes, Gautier, and Berthold, have sought in various ways to make it clearer, *e.g.* by regarding the reproduction of like by like as an instance of organic memory. As these suggestions are unessential to our argument, I shall merely notice that there are plenty of them.

How far has this early separation of the future reproductive cells from the developing body been observed? It has been observed in several worm-types—leeches, *Sagitta*, thread-worms, Polyzoa,—in some Arthropods (*e.g. Moina* among crustaceans, *Chironomus* among Insects, Phalangidæ among spiders), and with less distinctness in a number of other organisms, both animal and vegetable. In most of the higher animals, however, the future reproductive cells are not observable till development has proceeded for some days or weeks. To explain this difficulty, Weismann has elaborated a theory which he calls "*the continuity of the germ-plasma.*" The general idea of this theory is that of organic continuity between generations, and this Weismann has done momentous service in expounding. But for the detailed theory by which he seeks to overcome the difficulty which has been noticed above I refer those interested to Weismann's *Papers on Heredity* (Trans. Oxford, 1889).

4. **The Inheritance of Acquired Characters.**—(*a*) Historical.—We have seen that variations, or changes in character, may be *constitutional*, *i.e.* innate in the germ; or *functional*, *i.e.* due to use or disuse; or *environmental*, *i.e.* due to influences of nutrition and surroundings. Many naturalists have believed that gains or losses due to any of these three sources of change might be transmitted from parent to offspring. But nowadays the majority, with Profs. Weismann and Lankester at their head, deny the transmissibility of either functional or environmental changes, and believe that inborn, germinal, or constitutional variations alone are transmissible.

This scepticism is not strictly modern. The editor, whoever he was, of Aristotle's *Historia Animalium*, differed from his master as to the inheritance of injuries and the like. Kant maintained the non-inheritance of extrinsic variations, and Blumenbach cautiously inclined to the same negative position. In more recent times the veteran morphologist His expressed a strong conviction against the inheritance of acquired characters, and the not less renowned physiologist Pflüger is also among the sceptics. A few sentences from Galton (1875), whose far-sightedness has been insufficiently acknowledged, may be quoted: "The inheritance of characters acquired during the lifetime of the parents includes much questionable evidence, usually difficult of verification. We might almost reserve our belief that the structural cells can react on the sexual elements at all, and we may be confident that at the most they do so in a very faint degree—in other words, that acquired modifications are barely, if at all, *inherited* in the correct sense of that word."

But Weismann brought the discussion to a climax by altogether denying the transmissibility of acquired characters.

(*b*) Weismann's position.—Weismann's reasons for

maintaining that no acquired characters are transmissible are twofold,—first because the evidence in favour of such transmission consists of unverifiable anecdotes; second because the "germ-plasma," early set apart in the development of the body, remains intact and stable, unaffected by the vicissitudes which beset the body.

It is natural that Weismann, who realised so vividly the continuity between germ and germ, should emphasise the stability of the "germ-plasma," that he should regard it as leading a sort of charmed life within the organism unaffected by changes to which the body is subject. But has he not exaggerated this insulation and stability?

Of course Weismann does not deny that the body may exhibit functional and environmental variations, but he denies that these can spread from the body so as to affect the reproductive cells thereof, and unless they do so, they cannot be transmitted to the offspring.

On the other hand, innate or germinal characters must be transmitted. They crop up in the parent because they are involved in the fertilised egg-cell. But as the cell which gives rise to the offspring is by hypothesis similar to and more or less directly continuous with the cell which gave rise to the parent, similar constitutional variations will crop up in the offspring.

We must admit that most of the old evidence adduced in favour of the transmission of acquired characters may be called a "handful of anecdotes." For scepticism was undeveloped, and when a character acquired by a parent reappeared in the offspring, it was too readily regarded as transmitted, whereas it may often have been acquired by the offspring just as it was by the parent.

Weismann has two saving clauses, which make argument against his position peculiarly difficult. (1) He admits that the germ-plasma may be modified "ever so little" by changes of nutrition and growth in the body; but may not an accumulation of many "ever-so-littles" amount to the transmission of an acquired character? (2) He admits that external conditions, such as climate, may influence the reproductive cells *along with*, though not exactly *through*, the body; but this is a distinction too subtle to be verified.

These two saving-clauses seem to me to affect the stringency of Weismann's conclusion, but in his view they do not affect the main proposition that definite somatic modifications or changes in the body due to function or environment have no effect on the reproductive cells, and therefore no transmission to offspring.

(*c*) Arguments against Weismann's position.—In arguing against Weismann's position that no acquired characters are inherited, I shall first illustrate the arguments of others, and then emphasise that which appears to me at present most cogent.

(1) Some have cited against Weismann various cases where the effects of mutilation seemed to be transmitted, and Weismann has spent some time in experimenting with mice in order to see whether cutting off the tails for several generations did not eventually make the tails shorter. It did not—a result which might have been foretold. For we have known for many years that the mutilations inflicted on sheep and other domesticated animals had no measurable effect on the offspring. Even the numerous cases of tailless kittens produced from artificially curtailed cats have no cogency in face of the fact that tailless sports often arise from normal parents. Moreover, it is for many reasons not to be expected that the results of curtailment and the like should be inherited. For there is great power of regenerating lost parts even in the individual lifetime; the result of cutting off a tail is for most part merely a minus quantity to the organism; the imperfectly known physiological reaction on nerves and blood-vessels might perhaps result in a longer rather than a shorter tail in the offspring.

(2) Various pathologists, led by Virchow, have emphasised the fact that many diseases are inherited, but their arguments have usually shown how easy it is to misunderstand Weismann's position. No doubt many malformations and diseases reappear through successive generations, but there is lack of evidence to show that the pathological variations were not germinal to begin with. It is sadly

interesting to learn that colour-blindness has been known to occur in the males only of six successive generations, deaf mutism for three, finger malformations for six, and so with harelip and cleft palate, and with tendencies to consumption, cancer, gout, rheumatism, bleeding, and so on. But these facts do not prove the transmission of functional or environmental variations; they only corroborate what every one allows, that innate, congenital, constitutional characters

FIG. 70.—Half-lop rabbit, an abnormal variation, which by artificial selection has become constant in a breed. (From Darwin.)

tend to be transmitted. Yet some cases recently stated by Prof. Bertram Windle seem to suggest that some pathological conditions acquired by function may be transmitted. But even if a non-constitutional pathological state acquired by a parent reappeared in the offspring, we require to show that the offspring did not also acquire it by his work or from conditions of life, as his parent did before him.

(3) Some individual cases seem to stand some criticism.

Two botanists, Hoffmann and Detmer, have noted such facts as the following—scant nutrition influenced the flowers of poppy, *Nigella*, dead-nettle, and the result was transmitted; peculiar soil conditions altered the root of the carrot, and the result was transmitted.

Semper gives a few cases such as Schmankewitsch's transformation of one species of brine-shrimp (*Artemia*) into another, throughout a series of generations during which the salinity of the water was slowly altered.

Eimer has written a book of which even the title, "The Origin of Species, according to the laws of organic growth, through the inheritance of acquired characters," shows how strongly he supports the affirmative side of our question. But much as I admire and agree with many parts of Eimer's work, I do not think that all his examples of the inheritance of acquired characters are cogent. One of the strongest is that cereals from Scandinavian plains transplanted to the mountains become gradually accustomed to develop more rapidly and at a lower temperature, and that when returned to the plains they retain this power of rapid development. I am inclined to think that the strongest part of Eimer's argument is that in which he maintains that certain effects produced upon the nervous system by peculiar habits are transmissible.

(4) Another mode of argument may be considered. To what conception of evolution are we impelled if we deny the inheritance of acquired characters? Weismann believes that he has taken the ground from under the feet of Lamarckians and Buffonians, who believe in the inheritance of functional and environmental variations. The sole fount of change is to be found in the mingling of the kernels of two cells at the fertilisation of the ovum. On these variations natural selection works.

But even if we do not believe in the inheritance of acquired characters, it is open to us to maintain that by cumulative constitutional variations in definite directions species have grown out of one another in progressive evolution. Thus we are not forced to restrict our interpretations of the marvel and harmony of organic nature to the theory of the action of natural selection on indefinite fortuitous variations.

Prof. Ray Lankester's convictions on this subject are so

strong, and his dismissal of Lamarckian theory is so emphatic, that I shall select one of his illustrations by way of contrasting his theory with that of Lamarckians.

Many blind fishes and crustaceans are found in caves. Lamarckians assume, as yet with insufficient evidence, that the blindness is due to the darkness and to the disuse of the eyes. Changes thus produced are believed, again with insufficient evidence, to be transmitted and increased, generation after generation. This is a natural and simple theory, but it is not a certain conclusion.

What is Prof. Ray Lankester's explanation?

"The facts are fully explained by the theory of natural selection acting on congenital fortuitous variations. Many animals are born with distorted or defective eyes whose parents have not had their eyes submitted to any peculiar conditions. Supposing a number of some species of Arthropods or fish to be swept into a cavern, those individuals with perfect eyes would follow the glimmer of light and eventually escape to the outer air, leaving behind those with imperfect eyes to breed in the dark place. In every succeeding generation this would be the case, and even those with weak but still seeing eyes would in the course of time escape, until only a pure race of eyeless or blind animals would be left in the cavern." This is a possible explanation, but it is not a certain conclusion.

(5) The argument which I would urge most strongly is based on general physiological considerations. It gives no demonstration, but it seems to establish a presumption against Weismann's conclusion. He maintains that functional and environmental changes in the body cannot be transmitted because such changes cannot reach the stable and to some extent insulated reproductive elements. But this *cannot* requires proof, just as much as the converse *can*.

The organism is a unity; cell is often linked to cell by bridges of living matter; the blood is a common medium carrying food and waste; nervous relations bind the whole in harmony. Would it not be a physiological miracle if the reproductive cells led a charmed life unaffected even by influences which touch the very heart of the organism? Is it unreasonable to presume that *some* influences of habit and conditions, of training and control, saturate the organism thoroughly enough to affect every part of it?

A slight change of food affects the development of the reproductive organs in a bee-grub, and makes a queen out of what otherwise would have been a worker. A difference of diet causes a brood of tadpoles to become almost altogether female. There is no doubt that some somatic changes affect the reproductive cells in some way. Is it inconceivable that they affect them in such a precise way that bodily changes may be transmitted?

It must be admitted that it is at present impossible to give an explanation of the way in which a modification of the brain can affect the cells of the reproductive organs. The only connections that we know are by the blood, by nervous thrills, by protoplasmic continuity of cells. But there are many indubitable physiological influences which spread through the body of which we can give no rationale. Because we cannot tell how an influence spreads, we need not deny its existence.

It is at least conceivable that a deep functional or environmental change may result in chemical changes which spread from cell to cell, that characteristic products may be carried about by the blood and absorbed by the unspecialised reproductive cells, that nervous thrills of unknown efficacy may pass from part to part. Nor do we expect that more than a slight change will be transmitted in one generation.

Weismann traces all variations ultimately to the action of the environment on the original unicellular organisms. These are directly affected by surrounding influences, and as they have no "body" nor specialised reproductive elements, but are single cells, it is natural that the characters acquired by a parent-cell should also belong to the daughter-units into which it divides. And if so, is it not possible that the reproductive cells of higher animals, being equivalent to Protozoa, may be definitely affected by their immediate environment, the body? Moreover, if it were proved that the definite changes produced on an individual by influences of use, disuse, and surroundings, do not reach the reproductive cells, and cannot, therefore, be transmitted, it is not thereby proved that secondary results or some results of such definite changes may not have some effect on the germ-cells. The conditions are so complex that it seems rash to deny the possibility of such influence.

Certainly it is no easy task to explain all the adaptations to strange surroundings and habits, or the majority of animal instincts, or the progress of men, apart from the theory that some of the results of environmental influence and habitual experience are transmitted. I am certainly unable to reconcile myself to the opinion that the progress of life is due to the action of natural selection on fortuitous, indefinite, spontaneous variations.

I believe that the conclusion of the whole matter should be an emphatic "not proven" on either side, while the practical corollary is that we should cease to talk so much about possibilities (in regard to which one opinion is often as logically reasonable as another), and betake ourselves with energy to a study of the facts.

5. **Social and Ethical Aspects.**—All the important biological conclusions have a human interest.

The fact of organic continuity between germ and germ helps us to realise that the child is virtually as old as the parent, and that the main line of hereditary connection is not so much that between parent and child as "that between the sets of elements out of which the personal parents had been evolved, and the set out of which the personal child was evolved." "The main line," Galton says, "may be rudely likened to the chain of a necklace, and the personalities to pendants attached to the links." To this fact social inertia is largely due, for the organic stability secured by germinal continuity tends to hinder evolution by leaps and bounds either forwards or backwards. There is some resemblance between the formula of heredity and the first law of motion. The practical corollary is respect for a good stock.

That each parent contributes almost equally to the offspring suggests the two-sided responsibility of parentage; but the fact has to be corrected by Galton's statistical conclusion that the offspring inherits a fourth from each parent, and a sixteenth from each grandparent! Inherited capital is not merely dual, but multiple like a mosaic.

If we adopt a modified form of Weismann's conclusion, and believe that only the more deeply penetrating acquired characters are transmitted, we are saved from the despair suggested by the abnormal functions and environments of our civilisation.

And just in proportion as we doubt the transmission of desirable acquired characters, so much the more should we desire to secure that improved conditions of life foster the individual development of each successive generation.

That pathological conditions, innate or congenital in the organism, tend to be transmitted, suggests that men should be informed and educated as to the undesirability of parentage on the part of abnormal members of the community.

But while no one will gainsay the lessons to be drawn from the experience of past generations, it should be noticed that Virchow and others have hinted at an "optimism of pathology," since some of the less adequately known abnormal variations may be associated with new beginnings not without promise of possible utility. It seems, moreover, that by careful environment and function, or by the intercrossing of a slightly tainted and a relatively pure stock, a recuperative or counteractive influence may act so as to produce comparatively healthy offspring, thus illustrating what may be called "the forgiveness of nature."

6. **Social Inheritance.** — The widest problems of heredity are raised when we substitute "fraternities" for individuals, or make the transition to social inheritance—the relation between the successive generations of a society.

The most important pioneering work is that of Galton, whose unique papers have been recently summed up in a work entitled *Natural Inheritance*. Galton derived his data from his *Records of Family Faculties*, especially concerning stature, eye-colour, and artistic powers; and his work has been in great part an application of the statistical law of Frequency of Error to the records accumulated.

The main problem of his work is concerned with the strange regularity observed in the peculiarities of great populations throughout a series of generations. "The large do not always beget the large, nor the small the small; but yet the observed proportion between the large and the small, in each degree of size and in every quality hardly varies from one generation to another." A specific average is sustained. This is not because each individual leaves his like behind him, for this is not the case. It is rather due to the fact of a regular regression or deviation which brings the offspring of extraordinary parents in a definite ratio nearer the average of the stock.

"However paradoxical it may appear at first sight, it is theoretically a necessary fact, and one that is clearly confirmed by observation, that the stature of the adult offspring must on the whole be more mediocre than the stature of their parents—that is to say, more near to the median stature of the general population. Each peculiarity of a man is shared by his kinsmen, but *on an average* in a less degree. It is reduced to a definite fraction of its amount, quite independently of what its amount might be. The fraction differs in different orders of kinship, becoming smaller as they are more remote."

Yet it must not be supposed that the value of a good stock is under-estimated by Galton, for he shows how the offspring of two ordinary members of a gifted stock will not regress like the offspring of a couple equal in gifts to the former, but belonging to a poorer stock, above the average of which they have risen.

Yet the fact of regression tells against the full transmission of any signal talent. Children are not likely to differ from mediocrity so widely as their parents. "The more bountifully a parent is gifted by nature, the more rare will be his good fortune if he begets a son who is as richly endowed as himself, and still more so if he has a son who is endowed more largely." But "The law is even-handed; it levies an equal succession-tax on the transmission of badness as of goodness. If it discourages the extravagant hope of a gifted parent that his children will inherit all his powers, it no less discountenances extravagant fears that they will inherit all his weakness and disease."

The study of individual inheritance, **as** in Galton's *Hereditary Genius*, may tend to develop an aristocratic and justifiable pride of race when a gifted lineage is verifiable for generations. It may lead to despair if the records of family diseases be subjected to investigation.

But the study of social inheritance is at once more democratic and less pessimistic. The nation is a vast fraternity, with an average towards which the noble tend, but to which the offspring of the under-average as surely approximate. Measures which affect large numbers are thus more hopeful than those which artificially select a few.

Even when we are doubtful as to the degree in which acquired characters are transmissible, we cannot depreciate the effect on individuals of their work and surroundings. In fact there should be the more earnestness in our desire to conserve healthful function and stimulating environment of every kind, for these are not less important if their influences must needs be repeated on each fresh generation. "There was a child went forth every day; and the first object he looked upon, that object he became; and that object became part of him for the day, or a certain part of the day, or for many years, or for stretching cycles of years."[1]

Nor can we forget how much a plastic physical and mental education may do to counteract disadvantageous inherited qualities, or to strengthen characters which are useful.

Every one will allow at least that much requires to be done in educating public opinion, not only to recognise all the facts known in regard to heredity, but also to admit the value and necessity of the art which Mr. Galton calls "eugenics," or in frank English " good-breeding."

[1] Walt Whitman's "Assimilations."

APPENDIX I

ANIMAL LIFE AND OURS

A. Our Relation to Animals

1. **Affinities and Differences between Man and Monkeys.** In one of the works of Broca, a pioneer anthropologist of renown, there is an eloquent apology for those who find it useful to consider man's zoological relations.

"Pride," he says, "which is one of the most characteristic traits of our nature, has prevailed with many minds over the calm testimony of reason. Like the Roman emperors who, enervated by all their power, ended by denying their character as men, in fact, by believing themselves demigods, so the king of our planet pleases himself by imagining that the vile animal, subject to his caprices, cannot have anything in common with *his* peculiar nature. The proximity of the monkey vexes him, it is not enough to be king of animals; he wishes to separate himself from his subjects by a deep unfathomable abyss; and, turning his back upon the earth, he takes refuge with his menaced majesty in a nebulous sphere, 'the human kingdom.' But anatomy, like that slave who followed the conqueror's chariot crying, *Memento te hominem esse*, anatomy comes to trouble man in his naïve self-admiration, reminding him of the visible tangible facts which bind him to the animals."

Let us hearken to this slave a little, remembering Pascal's maxims: "It is dangerous to show man too plainly how like he is to the animals, without, at the same time, reminding him of his greatness. It is equally unwise to impress him with his greatness, and not with his lowliness. It is worse to leave him in ignorance of both. But it is very profitable to recognise the two facts."

It is many years since Owen—now a veteran among anatomists —described the "all-pervading similitude of structure" between man and the highest monkeys. Subsequent research has continued to add corroborating details. As far as structure is concerned, there is much less difference between man and the gorilla than between the gorilla and a monkey like a marmoset. Yet differences between man and the anthropoid apes do exist. Thus man alone is thoroughly erect after his infancy is past, his head weighted with a heavy brain does not droop forward, and with his erect attitude his perfect development of vocal mechanism is perhaps connected. We plant the soles of our feet flat on the ground, our great toes are usually in a line with the rest, and we have better heels than monkeys have, but no emphasis can be laid on the old distinction which separated two-handed men (Bimana) from the four-handed monkeys (Quadrumana), nor on the fact that man is peculiarly naked. We have a bigger forehead, a less protrusive face, smaller cheek-bones and eyebrow ridges, a true chin, and more uniform teeth than the anthropoid apes. More important, however, is the fact that the weight of the gorilla's brain bears to that of the smallest brain of an adult man the ratio of 2 : 3, and to the largest human brain the ratio of 1 : 3; in other words, a man may have a brain three times as heavy as that of a gorilla. The brain of a healthy human adult never weighs less than 31 or 32 ounces; the average human brain weighs 48 or 49 ounces; the heaviest gorilla brain does not exceed 20 ounces. "The cranial capacity is never less than 55 cubic inches in any normal human subject, while in the orang and the chimpanzee it is but 26 and 27½ cubic inches respectively."

But differences which can be measured and weighed give us little hint of the characteristically human powers of building up ideas and of cherishing ideals. It is not merely that man profits by his experience, as many animals do, but that he makes some kind of theory of it. It is not merely that he works for ends which are remote, as do birds and beavers, but that he controls his life according to conscious ideals of conduct. But I need not say much in regard to the characteristics of human personality, we are all conscious of them, though we may differ as to the words in which they may be expressed; nor need I talk about man's power of articulate speech, nor his realisation of history, nor his inherent social sympathies, nor his gentleness. For all recognise that the higher life of men has a loftier pitch than that of animals, while many think that the difference is in kind, not merely in degree.

2. **Descent of Man.**—The arguments by which Darwin and others have sought to show that man arose from an ancestral type common to him and to the higher apes are the same as those used to substantiate the general doctrine of descent. For the *Descent of Man* was but the expansion of a chapter in the *Origin of Species*;

the arguments used to prove the origin of animal from animal were adapted to rationalise the ascent of man.

(*a*) *Physiological.*—The bodily life of man is like that of monkeys; both are subject to the same diseases; various human traits, such as gestures and expressions, are paralleled among the "brutes"; and children born during famine or in disease are often sadly ape-like.

(*b*) *Morphological.*—The structure of man is like that of the anthropoid apes, none of his distinctive characters except that of a heavy brain being momentous, and there are about seventy vestigial structures in the muscular, skeletal, and other systems.

(*c*) *Historical.*—There is little certainty in regard to the fossil remains of prehistoric man, but some of these suggest more primitive skulls, while the facts known about ancient life show at least that there has been progress along certain lines. Moreover, there is the progress of each individual life, from the apparently simple egg-cell to the minute embryo, which is fashioned within the womb into the likeness of a child, and being born grows from stage to stage, all in a manner which it is hard to understand if man be not the outcome of a natural evolution.

3. **Various Opinions about the Descent of Man.**—But opinion in regard to the origin of man is by no means unanimous.

(*a*) A few authorities, notably A. de Quatrefages, maintain a conservative position, believing that the evolutionist's case has not been sufficiently demonstrated. But the majority of naturalists believe the reverse, and think that the insufficiencies of evidence in regard to man are counterbalanced by the force of the argument from analogy.

(*b*) Alfred Russel Wallace has consistently maintained a position which seems to many a very strong one. "I fully accept," he says, "Mr. Darwin's conclusion as to the essential identity of man's bodily structure with that of the higher mammalia, and his descent from some ancestral form common to man and the anthropoid apes. The evidence of such descent appears to me overwhelming and conclusive. Again, as to the cause and method of such descent and modification, we may admit, at all events provisionally, that the laws of variation and natural selection, acting through the struggle for existence and the continual need of more perfect adaptation to the physical and biological environments, may have brought about, first that perfection of bodily structure in which he is so far above all other animals, and in co-ordination with it the larger and more developed brain, by means of which he has been able to utilise that structure in the more and more complete subjection of the whole animal and vegetable kingdoms to his service."

"But because man's physical structure has been developed from an animal form by natural selection, it does not necessarily follow that his mental nature, even though developed *pari passu* with it, has been developed by the same causes only." Wallace then goes on to show that man's mathematical, musical, artistic, and other higher faculties could not be developed by variation and natural selection alone. "Therefore some other influence, law, or agency is required to account for them." Indeed this unknown cause or power may have had a much wider influence, extending to the whole course of his development. "The love of truth, the delight in beauty, the passion for justice, and the thrill of exultation with which we hear of any act of courageous self-sacrifice, are the workings within us of a higher nature which has not been developed by means of the struggle for material existence." At the origin of living things, at the introduction of consciousness, in the development of man's higher faculties, "a change in essential nature (due, probably, to causes of a higher order than those of the material universe) took place." "The progressive manifestations of life in the vegetable, the animal, and man—which we may classify as unconscious, conscious, and intellectual life—probably depend upon different degrees of spiritual influx."

In discussing problems such as this there is apt to be misunderstanding, for words are "but feeble light on the depth of the unspoken," and perhaps no man appreciates his brother's philosophy. Therefore, I refrain from seeking to controvert what Wallace has said, especially as I also believe that the nature of life and mind are secrets to us all, and that the higher life of man cannot be explained by indefinite variations which happened to prosper in the course of natural selection.

But it seems to me (1) to be difficult to divide man's self into an animal nature which has been naturally evolved and "a spiritual nature which has been superadded," or to separate man's higher life from that of some of the beasts. (2) When we find that any fact in our experience, such as human reason, cannot be explained on the theory of evolution which we have adopted, it does not follow that the reality in question has not been naturally evolved, it only follows that our theory of evolution is imperfect. A theory is not proved to be complete because it explains many facts, but it is proved to be incomplete if it fails to explain any. Thus if man's higher nature cannot be explained by the theory of natural selection in the struggle for existence, then that theory is incomplete, but there may be other theories of evolution which are sufficient. (3) It is difficult to know what is meant by spiritual influx—for our opinions in regard to those matters vary with individual experience. We may mean to suggest the interpolation of a power of a secret and supersensory nature, distinct from that power which is everywhere present in sunbeam and raindrop, bird and flower. Then we are abandoning the theory of a continuous natural evolution. Or we may mean to suggest that when life and mind and man began to be, then possibilities of action and reaction hitherto latent became real, and all things became in a sense new. Then, while maintaining that life and mind are new realities with new powers, we are still consistent believers in a continuous natural evolution. (4) Perhaps the simplest conception is that more than once suggested in this book, that the world is one not twofold, that the spiritual influx is the primal reality, that there is nothing in the end which was not also in the beginning.

(*c*) Prof. Calderwood has recently stated with clearness and conciseness what difficulties surround the task of those who would explain the evolution of man. "So far as the human organism is concerned, there seem no overwhelming obstacles to be encountered by an evolution theory; but it seems impossible under such a theory to account for the appearance of *homo sapiens*—the thinking, self-regulating life, distinctively human." Again, I have no desire to enter into controversy, for I recognise the difficulties which the student of comparative psychology must tackle, but it seems important that the following consideration should be kept in mind.

It is not the first business of the evolutionist to find out how one reality has grown out of another, but to marshal the arguments which lead him to conclude that one reality *has* so evolved. We have only a vague idea how a backbone arose, but that need not hinder us from believing that backboned animals were evolved from backboneless if there be sufficient evidence in favour of this conclusion. We do not know how birds arose from a reptile stock, but that they did so arise is fairly certain. We cannot explain the intelligence of man in terms of the activity of the brain; we are equally at a loss in regard to the intelligence of an ant. What we have to do is to compare the structure of man's brain with that of the nearest animals, and the nature of human intelligence with that of the closest approximations, drawing from the results of our comparison what conclusion we can. The general doctrine of descent may be established independently of the investigations of physiologist and psychologist, valuable as these may be in elucidating the way in which the great steps of progress have been made.

(*d*) Finally there is the opinion of many that man is altogether too marvellous a being to have arisen from any humbler form of life. But to others this ascent seems the stamp of man's nobility.

4. **Ancestors of Man.**—Of these we know nothing. The anthropoid apes approach him most closely, each in some particular respect, but none of them nor any known form of life can be called

man's ancestor. It is possible that the race of men—for of a first man evolutionists cannot speak—began in Miocene times, as offshoots from an ancestral stock common to them and to the anthropoids. We often hear of " the missing link," but surely no one expects to find him alive. And while we have still much to learn from the imperfect geological record, it must be remembered that what most distinguishes man will not be remarkable in a fossil, for brains do not petrify except metaphorically, nor can we look for fossilised intelligence or gentleness.

FIG. 71.—Young gorilla. (From Du Chaillu.)

5. **Possible Factors in the Ascent of Man.**—In regard to the factors which secured man's ascent from a humbler form of life we can only speculate.

(*a*) We have already explained that organisms vary, that the offspring differ from their parents, that the more favourable changes prosper, and that the less fit die out of the struggle. Thus the race is lifted. Now, from what we know of men and monkeys, it seems likely that in the struggles of primitive man cunning was more important than strength, and if intelligence now became, more than ever before, the condition of life or death, wits would tend to develop rapidly.

(*b*) When habits of using sticks and stones, of building shelters, of living in families, began—and some monkeys exhibit these—it is likely that wits would increase by leaps and bounds.

(*c*) Professor Fiske and others have emphasised the importance of prolonged infancy, and this must surely have helped to evolve the gentleness of mankind.

(*d*) Among many monkeys society has begun. Families combine for protection, and the combination favours the development both of emotional and intellectual strength. Surely "man did not make society, society made man."

B. *Our Relation to Biology.*

6. The Utility of Science.—As life is short, all too short for learning the art of living, it is well that we should criticise our activities, and favour those which seem to yield most return of health and wealth and wisdom.

We are so curious about all kinds of things, so omnivorously hungry for information, that the most trivial department of knowledge or science may afford exercise and mental satisfaction to its votaries. The interest and pleasantness of science is therefore no criterion.

Nor can we be satisfied with the assertion that science should be pursued for science's sake. As in regard to the kindred dictum, "art for art's sake," we require further explanation—some ideal of science and art. For it is not evident that knowledge is a good in itself, especially if that knowledge be gained at the expense of the emotional wealth which is often associated with healthy ignorance.

Nor is it safe to judge scientific activity by the material results which the application of knowledge to action may yield. For a seed of knowledge may lie dormant for centuries before it sends its shoots into life, and many of the material results of applied science are not unmixed blessings. Moreover, too narrow a view may be taken of material results, so-called " necessaries " of existence may be exalted over the "super-necessaries" essential to life; in short, what lies about the mouth—the nose, the ears, the eyes, the brain—may be forgotten.

We are nearer the truth if we combine the different standards of science, and unify them by reference to the human ideal.[1] The utility of science, and of biology among the other kinds of knowledge, is to supply a basis of fact—

(*a*) For the practice of useful arts (such as hygiene and education), and for the guidance of conduct :
(*b*) For the satisfaction of our desire to understand and enjoy the world and our life in it.

7. Practical Justification of Biology.—The world of life is so web-like that almost any part may touch or thrill us. It is therefore well that we should learn what we can about it.

On plants we are very dependent for food and drink, for shelter

---

[1] See Ruskin, *The Eagle's Nest* (1880).

and clothing, and for delight. Their evil influence is almost restricted to that of disease germs and poisonous herbs.

Animals likewise furnish food (perhaps to an unwholesome extent); and parts of their bodies are used (sometimes carelessly) in manifold ways. Among those which are domesticated, some, such as canary and parrot, cat and dog, are kept for the pleasure they give to many; others, such as dog, horse, elephant, and falcon, are used in the chase; others, notably the dog, assist in shepherding; horse and ass, reindeer and cattle, camel and elephant, are beasts of burden; others yield useful products, the milk of cows and goats, the eggs of birds, the silk of silkworms, and the honey of bees.

Formerly of much greater importance for good and ill as direct rivals, animals have, through man's increasing mastery of life, become less dangerous and more directly useful. Only in primitive conditions of life and in thinly-peopled territories is something of the old struggle still experienced. Their influence for ill is now for the most part indirect,—on crops and stocks. Parasites are common enough, but rarely fatal. The serpent, however, still bites the heel of progressive man.

Man's relations with living creatures are so close that systematic knowledge about them is evidently of direct use. Indeed it is in practical lore that both botany and zoology have their primal roots, and from these, now much strengthened, impulses do not cease to give new life to science.

If increase of food-supply be desirable, biology has something to say about soil and cereals, about fisheries and oyster-culture. The art of agriculture and breeding has been influenced not a little by scientific advice, though much more by unrationalised experience. If wine be wanted, the biologist has something to say about grafting and the *Phylloxera*, about mildew and Bacteria. It is enough to point to the succession of discoveries by which Pasteur alone has enriched science and benefited humanity.

But if we take higher ground and consider as an ideal the healthfulness of men, which is one of the most obvious and useful standards of individual and social conduct, the practical justification of biological science becomes even more apparent.

Medicine, hygiene, physical education, and good-breeding (or "eugenics") are the arts which correspond to the science of biology, just as education is applied psychology, as government is applied sociology, and as many industries are applied chemistry and physics. It would be historically untrue to say that the progress in these arts was due to progress in the parallel sciences; in fact the progressive impulse has often been from art to science. "La pratique a partout devancé la théorie," Espinas says, and all historians of science would in the main confirm this. But it is also true that science reacts on the arts and sometimes improves them.

There may be peculiar aberrations of the art of medicine due to the progress of the science thereof, but these are because the science is partial, and hardly affect the general fact that scientific progress has advanced the art of healing. The results of science have likewise supplied a basis to the endeavours to prevent disease and to increase healthfulness, not only by definite hygienic practice but perhaps still more by diffusing some precise knowledge of the conditions of health.

The generalisations of biology, realised in men's minds, must in some measure affect practice and public opinion. Spencer's induction that the rate of reproduction varies inversely with the degree of development sheds a hopeful light on the population question; the recognition of the influence which function and surroundings have upon the organism suggests criticism of many modes of economic production; a knowledge of the facts and theory of heredity must have an increasing influence on the art of eugenics. Nor can I believe that the theory of evolution which men hold, granting that it is in part an expression of their life and social environment, does not also react on these.

In short, the direct application of biological knowledge in the various arts of medicine, hygiene, physical education, and eugenics, helps us to perfect our environment and our relations with it, helps us to discover—if not the "elixir vitæ"—some not despicable substitute. And likewise, a realisation of the facts and principles of biology helps us to criticise, justify, and regulate conduct, suggesting how the art of life may be better learned, how human relations may be more wisely harmonised, how we may guide and help the ascent of man.

8. **Intellectual Justification of Biology.**—But another partial justification of Biology is found in our desire to understand things, in our dislike of obscurities, in our inborn curiosity. There is an intellectual as well as a practical and ethical justification of the study of organic life.

Through our senses we become aware of the world of which we form a part. We cannot know it in itself, for we are part of it and only know it as it becomes part of us. We know only fractions of reality—real at least to us—and these are unified in our experience.

(1) In the world around us we are accustomed to distinguish four orders of facts. "Matter" and "energy" we call those which seem to us fundamental, because all that we know by our senses are forms of these. The study of matter and energy—or perhaps we may say the study of matter in motion—considered apart from

life, we call Physics and Chemistry, of which astronomy, geology, etc., are special departments.

(2) But we also know something about plants and animals, and while all that we know about them is still dependent upon changes of matter and motion, yet we recognise that the activities of the organism cannot at present be expressed in terms of these. Therefore we find it convenient to speak of life as a new reality, while believing that it is the result of some combination of matters and energies, the secret of which is hidden.

(3) But we are also aware of another reality, our own mind. Of this we have direct consciousness and greater certainty than about anything else. And while some would say that what we are conscious of when we think is a protoplasmic change in our brain cells or is a subtle kind of motion, it is truer to say that we are conscious of ourselves. It is our thought that we know, it is our feeling that we feel, and as we cannot explain the thought or the feeling in terms of protoplasm or of motion, we find it convenient to speak of mind as a new reality, while believing it to be essentially associated with some complex activity of protoplasm the secret of which is hidden. For our knowledge of our own mental processes, and of those inferred to be similar in our fellows, and of those inferred to be not very different in intelligent animals, we establish another science of Psychology.

(4) But we also know something about the life of the human society of which we form a part. We recognise that it has a unity of its own, and that its activities are more than those of its individual members added up. We find it convenient to regard society as another synthesis or unity—though less definite than either organism or mind—and to our knowledge of the life and growth of society as a whole, we apply the term sociology.

Thus we recognise four orders of facts and four great sciences—

4. Society . . . . . . Sociology.
3. Mind . . . . Psychology.
2. Life . . . . Biology.
1. Matter and Energy . . . Physics and Chemistry.

Each of these sciences is dependent upon its predecessor. The student of organisms requires help from the student of chemistry and physics; mind cannot be discussed apart from body; nor can society be studied apart from the minds of its component members.

Each order of realities we may regard as a subtle synthesis of those which we call simpler. Life is a secret synthesis of matter and energy; mind is a subtle form of life; society is a unity of minds.

But it must be clearly recognised that the "matter and energy" which we regard as the fundamental realities are only known to us through what is for us the supreme reality—ourselves—mind. And as in our brain activity we know matter and energy as thought, I have adopted throughout this book what may be called a monistic philosophy.

Having recognised the central position of Biology among the other sciences, we have still to inquire what its task precisely is.

Our scientific data are (1) the impressions which we gather through our senses about living creatures, and (2) the deductions which we directly draw in regard to these. Our scientific aim is to arrange these data so that we may have a mental picture of the life around us, so that we may be better able to understand what that life is, and how it has come to be what it seems to be. Pursuing what are called scientific methods, we try to make the world of life and our life as organisms as intelligible as possible. We seek to remove obscurities of perception, to make the world translucent, to make a working thought-model of the world.

But we are apt to forget how ignorant we are about the realities themselves, for all the time we are dealing not with realities, but with impressions of realities, and with inferences from these impressions. On the other hand, we are apt to forget that our deep desire is not merely to know, but to enjoy the world, that the heart of things is not so much known by the man as it is felt by the child.

APPENDIX II

SOME OF THE "BEST BOOKS" ON ANIMAL LIFE

To recommend the "best books" on any subject is apt to be like prescribing the "best diet." Both depend upon age, constitution, and opportunities. The best book for me is that which does me most good, but it may be tedious reading for you. Moreover, books are often good for one purpose and not for another; that which helps us to realise the beauty and marvel of animal life may be of little service to those who are preparing for any of the numerous examinations in science. But the greatest difficulty is that we are often too much influenced by contemporary opinion, so that we lose our power of appreciating intellectual perspective.

The best way to begin the study of Natural History is to observe animal life, but the next best way is to read such accounts of observation and travel as are to be found in the works of Gilbert White, Thoreau, Richard Jefferies, and John Burroughs, or in Bates's *Naturalist on the Amazons*, Belt's *Naturalist in Nicaragua*, and Darwin's *Voyage of the "Beagle."* Sooner or later the student will seek more systematic books, but it is not natural that he should begin with a text-book of elementary biology.

In introducing you to the literature devoted to the study of animals, I shall avoid the bias of current opinion by following the history of zoology. I shall first name some of the more technical books; secondly, some of the more popular; thirdly, some of the more theoretical. If I may make the distinction, I shall first mention books on zoology, secondly those on natural history, thirdly those on biology.

A. Zoology.

(1) We can form a vivid conception of the history of zoology by comparing it with our own. In our childhood we knew and

cared more about the useful, dangerous, and strange animals than about those which were humble and familiar; we had more interest in haunts and habits than in structure and history; we were content with rough-and-ready classification, and cherished a feeling of superstitious awe in regard to the indistinctly-known forms of life. We were inquisitive rather than critical; we accepted almost any explanation of facts, and, if we tried to interpret, forced our borrowed opinions upon nature instead of trying to study things for ourselves. So was it with those naturalists who lived before Aristotle.

We must also recognise that the science of zoology had its beginnings in a practical acquaintance with animals, just as botany sprang from the knowledge of ancient agriculturists and herb-gatherers. Much information in regard to the earliest zoological knowledge has been gathered from researches into the history of words, art, and religious customs, and there is still much to be gleaned. Therefore I should recommend the student to dip into those books which discuss the early history of man, such as Lubbock's *Prehistoric Times* (1865), and *Origin of Civilisation* (1870); Tylor's *Primitive Culture* (1871), and *Anthropology* (1881); Andrew Lang's *Myths, Ritual, and Religion*; besides works on the history of philosophy, such as those of Schwegler and of Zeller, which give some account of ancient cosmogonies.

(2) But just as there are precocious children, so there was an early naturalist, whose works form the most colossal monument to the intellectual prowess of any one thinker. The foundations of zoology were laid by Aristotle, who lived 384-322 B.C. He collected many observations, and argued from them to general statements. He records over five hundred animals, and describes the structure and habits, the struggles and friendliness, of some of these. His is the first definite classification. His work was dominated by the idea that animal life is a unity and part of a larger system of things. In part his works should be read, and besides the great edition by Bekker (Berlin, 1831-40), there is a translation of *The Parts of Animals* by Dr. Ogle, and of *The History of Animals* by R. Cresswell. See also G. J. Romanes's "Aristotle as a Naturalist," *Nineteenth Century* (Feb. 1891, pp. 275-289).

(3) After the freedom of early childhood, and in most cases after precocity too, there comes a lull of inquisitiveness. Other affairs, practical tasks, games and combats, engross the attention, and parents sigh over dormant intellects; so the historian of zoology sighs over the fifteen centuries during which science slumbered. The foundations which Aristotle had firmly laid remained, but the walls of the temple of knowledge did not rise.

The seeds which he had sown were alive, but they did not germinate. Men were otherwise occupied, with practical affairs, with the tasks of civilisation alike in peace and war, though some at their leisure played with ideas which they did not verify. There were some exceptions; such as Pliny (23-79 A.D.), a diligent but uncritical collector of facts, and Galen (130-200 A.D.), a medical anatomist, who had the courage to dissect monkeys; besides the Spanish bishop Isidor in the seventh century, and various Arabian inquirers. It will not be unprofitable to look into the *Natural History* of Pliny, which has been translated by Bostock and Riley.

(4) But just as there is in our life a stage—happy are those who prolong it—during which we delight in fables and fairy tales, so there was a long period of mythological zoology. The schoolboy who puts horse-hairs into the brook, and returns after many days to find them eel-like worms, is doing what they did in the Middle Ages. For then fact and fiction were strangely jumbled; credulity ran riot along the paths of science; allegorical interpretations and superstitious symbolisms were abundant as the fancies which flit through the minds of dreamers. Scientific inquiry was not encouraged by the theological mood of the time; and just as Scotch children cherish *The Beasts of the Bible* as a pleasantly secular book with a spice of sacredness which makes it legitimate reading on the Sabbath, so many a mediæval naturalist had to cloak his observations in a semi-theological style.

In illustration of the mood of the mediæval naturalists, which is by no means to be carelessly laughed at, read John Ashton's *Curious Creatures* (Lond., 1890), in which much old lore is retold, often in the words of the original writers. The most characteristic expression of mythical Zoology is a production often called *Physiologus*. It is found in about a dozen languages and in many different forms, being in part merely a precipitate of floating traditions. It is partly like a natural history of the beasts of the Bible and prototype of many similar works, partly an account of the habits of animals, the study of which modern zoologists are apt to neglect, partly a collection of natural history fables and anecdotes, partly a treatise on symbolism and suggestive of the poetical side of zoology, partly an account of the medicinal and magical uses of animals. For many centuries it seems to have served as a text-book, a fact in itself an index to the slow progress of the science. Its influence on art and literature has been considerable, and it well illustrates the attempt to secure for the unextinguishable interest in living things a sanction and foothold under the patronage of theology. A series of fifty emblems is described, among others the lion which sleeps with its eyes open, the lizard which recovers its sight by looking at the sun, the eagle which renews its youth, the tortoise mistaken for an island, the serpent afraid of naked man, and the most miserable ant-lion, which is not able either to take one kind of food or digest the other.

(5) But delight in romance is replaced by a feeling of the need for definite knowledge, and the earlier years of adolescent manhood and womanhood are often very markedly characterised by a thirst and hunger for information. Which of us—now perhaps *blasé* with too much learning—does not recall the enthusiasm for knowing which once swayed our minds? Stimulated in a hundred ways by new experiences and responsibilities, our appetite for facts was once enormous. This was the mood of naturalists during the next great period in the history of zoology.

The freer circulation of men and thoughts associated with the Crusades; the discovery of new lands by travellers like Marco Polo and Columbus; the founding of universities and learned societies; the establishment of museums and botanic gardens; the invention of printing and the reappearance of Aristotle's works in dilution and translation; and many other practical, emotional, and intellectual movements gave fresh force to science, and indeed to the whole life of man. If we pass over some connecting links, such as Albertus Magnus in the thirteeenth century, we may call the period of gradual scientific renaissance that of the Encyclopædists. This somewhat cumbrous title suggests the omnivorous habits of those early workers. They were painstaking collectors of all information about all animals; but their appetite was greater than their digestion, and the progress of science was in quantity rather than in quality. Prominent among them were these four, the Englishman Edward Wotton (1492-1555), who wrote a treatise *De Differentiis Animalium*; the Swiss Conrad Gesner (1516-65), author of a well-known *Historia Animalium*; the Italian Aldrovandi (b. 1522); and the Scotsman Johnston (b. 1603).

About the middle of the eighteenth century the best aims of the Encyclopædists were realised in Buffon's *Histoire Naturelle*, which appeared in fifteen volumes between 1749 and 1767. This work not only describes beasts and birds, the earth and man, with an eloquent enthusiasm which was natural to the author and pleasing to his contemporaries, but is the first noteworthy attempt to expound the history or evolution of animals. Its range was very wide; and its successors are not so much single books as many different kinds of books, on geology and physical geography, on classification and physiology, on anthropology and natural history. There is a good French edition of Buffon's complete works by

A. Richard 1825-28), and at least one English translation. Three large modern books on natural history correspond in some degree to the *Histoire Naturelle*, viz. *Cassell's Natural History*, edited by P. Martin Duncan (6 vols.; London, 1882); *The Standard* or *Riverside Natural History*, edited by J. S. Kingsley (6 vols.; London, 1888); and a remarkable work well known as Brehm's *Thierleben*, of which a new (3rd) edition is at present in progress (10 vols.; Leipzig and Wien, 1890). Those who read German will find in Carus Sterne's (Ernst Krause's) *Werden und Vergehen* (3rd ed.; Berlin, 1886) the most successful attempt hitherto made to combine in one volume a history of the earth and its inhabitants.

(6) From Buffon till now the history of biology shows a progressive analysis, a deeper and deeper penetration into the structure and life of organisms. From external form to the internal organs, from organs to the tissues which compose them, from tissues to their elementary units or cells, and from cells to the living matter itself, has been the progress of the science of structure—*Morphology*. From habit and temperament to the work of organs, from the functions of organs to the properties of tissues, from these to the activities of cells, and from these finally to the chemical and physical changes in the living matter or protoplasm, has been the progress of the science of function—*Physiology*. Such is the lucid account which Prof. Geddes has given of the last hundred years' progress; see his article "Biology" in the new edition of Chambers's *Encyclopædia*. Following the metaphor on which we have already insisted, we may compare this century of analysis to the period of ordered and more intense study which in the individual life succeeds the abandonment of encyclopædic ambitions.

We should clearly understand the history of this gradually deepening analysis of animals; for if we would be naturalists' we must retread the same path. The history of biology has still to be written, but there are already some useful books and papers, notably—J. V. Carus, *Geschichte der Zoologie* (München, 1872); J. Sachs, *Geschichte der Botanik* (München, 1875), translated into English (Oxford, 1890); W. Whewell, *History of the Inductive Sciences* (London, 1840); articles "Morphology" and "Physiology," *Encyclopædia Britannica*, by P. Geddes and M. Foster; H. A. Nicholson, *Natural History: its Rise and Progress in Britain* (Edinburgh, 1888); A. B. Buckley, *Short History of Natural Science*; E. Perrier, *La Philosophie Zoologique avant Darwin* (Paris, 1884); Ernst Krause (Carus Sterne), *Die Allgemeine Weltanschauung in ihrer historischen Entwickelung* (Stuttgart, 1889). Very instructive, not least so in contrast, are two articles, "Biology" (in Chambers's *Encyclopædia*), by

P. Geddes, and "Zoology" (in *Encyclopædia Britannica*), by E. Ray Lankester.

If we think over the sketch which Professor Geddes has given, we shall see how easy it is to arrange the literature—the first step towards mastering it. (*a*) The early anatomists were chiefly occupied with the study of external and general features, very largely moreover with the purpose of establishing a classification. The *Systema Naturæ* of Linnæus (1st ed., 1735; 12th, 1768) is the typical work on this heavily-laden shelf of the zoological library. It is to such books that we turn when we wish to identify some animal, but the shelf is very long and most of the volumes are very heavy. Each chapter of Linné's *Systema* has been expanded into a series of volumes, or into some gigantic monograph like those included in the series of "*Challenger" Reports*, or *The Fauna and Flora of the Gulf of Naples*. If I am asked to recommend a volume from which the eager student may identify some British flower, I can at once place Hooker's *Flora* in his hands. But it is more difficult to help him to a work by which he may identify his animal prize. There are special works on British Mammals, Birds, Fishes, Molluscs, Insects, etc., but a compact *British Fauna* is much wanted. I shall simply mention Bronn's *Klassen und Ordnungen des Thierreiches*, a series of volumes still in progress; Leunis, *Synopsis des Thierreiches* (Hanover, 1886); the British Museum Catalogues (in progress); and P. H. Gosse's *Manual of Marine Zoology of the British Islands* (1856).

(*b*) Cuvier's *Règne Animal* (1829) is the typical book on the next plane of research—that concerned with the anatomy of organs. I should recommend the student on this path to begin with Professor F. Jeffrey Bell's *Comparative Anatomy and Physiology* (Lond., 1886); after which he will more readily appreciate the text-books on Comparative Anatomy by Huxley, Gegenbaur, Claus, Wiedersheim, Lang, etc. As an introduction I may also mention my *Outlines of Zoology* (Edin., 1892). As a book of reference Hatchett Jackson's edition of Rolleston's *Forms of Animal Life* (Oxford, 1888) is of great value, not least on account of its scholarly references to the literature of zoology. The zoological articles in the *Encyclopædia Britannica*, many of which are published separately, are not less useful. As guides in serious practical work may be noticed—*A Course of Elementary Instruction in Practical Biology* by Profs. T. H. Huxley and H. N. Martin, revised by Profs. G. B. Howes and D. H. Scott (Lond., 1888); Howes's *Atlas of Practical Elementary Biology* (Lond., 1885); *A Course of Practical Zoology* by Prof. A. Milnes Marshall and Dr. C. H. Hurst (3rd ed., Lond., 1892); Prof. C. Lloyd Morgan's *Animal Biology* (Lond., 1889); Vogt and Yung, *Traité d'Anatomie comparée pratique* (Paris, 1885-92) or in German (Braunschweig); Prof. W. K. Brooks's *Handbook of Invertebrate Zoology for Laboratories and Seaside Work* (Boston, 1882); Prof. T. J. Parker's *Zootomy* (Lond., 1884) and *Practical Biology* (Lond., 1891).

(*c*) As early as 1801, Bichât had **penetrated** beneath the organs to the tissues which compose them, and his *Anatomie Générale* is the forerunner of many works on minute anatomy or histology. From the comparative histology of animals by Leydig (*Histologie*, 1867) the zoological student must begin, but to follow it up he must have recourse to the pages of scientific journals. As a guide in microscopic work, Dr. Dallinger's new edition of Carpenter's well-known work, *The Microscope* (Lond., 1891) may be cited.

(*d*) In 1838-39, Schwann and Schleiden, two German naturalists, clearly stated a doctrine towards which investigation had been gradually tending, namely, that each organism was built up of cells, and originated from a fertilised egg-cell. In the establishment of this "cell-theory" the study of structure became deeper, and the investigation of animal cells still becomes more and more intense. To gain an appreciation of this step in analysis, the student may well begin with the article "Cell" in the new edition of Chambers's *Encyclopædia*, and with the articles "Morphology" and "Protozoa" in the *Encyclopædia Britannica*. From these he will discover how his studies may be deepened.

(*e*) Finally, with the improvement of microscopic instruments and technique, investigation has touched the bottom, as far as biology is concerned, in the study of the living stuff or protoplasm itself. Again, I refer you to the articles "Protoplasm" in the *Encyclopædia Britannica* and in Chambers's *Encyclopædia*.

I shall not follow the history of physiology in detail, but content myself with saying that (*a*) from the conception of a living body ruled by spirits or dominated by a temperament, physiologists passed to consider it (*b*) as an engine of living organs, then (*c*) as a complex web of tissues, then (*d*) as a city of cells, and finally (*e*) as a whirlpool of living matter. I recommend you to read first the article "Physiology" in the *Encyclopædia Britannica*, then Huxley's *Crayfish* (International Science Series), and his *Elementary Textbook of Physiology*, then Jeffrey Bell's *Comparative Anatomy and Physiology* and Lloyd Morgan's *Animal Biology*, after which you may pass to larger works such as the text-books of Kirkes (new ed., 1892); Bunge (Lond., 1890); Landois and Stirling, McKendrick, and Foster, and to the studies on comparative physiology by Krukenberg, *Vergleichend-Physiologische Studien* and *Vorträge* (Heidelberg, 1882-88).

In the above summary **nothing has been said about** the history of animals in their individual life (embryology), nor of their gradual appearance upon the earth (palæontology), nor about their distribution in space. As regards embryology, begin with the article in Chambers's *Encyclopædia*, and pass thence to the text-books of A. C. Haddon, F. M. Balfour, M. Foster and F. M. Balfour, O. Hertwig, Heider and Korschelt, etc. A short account of distribution in time will be found in A. Heilprin's *Distribution of Animals* (International Science Series), from which advanced students may pass to the *Text-book of Palæontology* by H. A. Nicholson and R. Lydekker (2 vols., Lond. and Edin., 1889), to the French work of Gaudry, *Les enchaînements du monde animal dans les temps géologiques* (Paris, 1888-90), or to the German works of Zittel and of Neumayr. Heilprin's book is again the best introduction to the study of distribution in space, while Wallace's *Geographical Distribution of Animals* (Lond., 1876) remains the principal work of reference.

For progressive research I may refer the student to the *Journal of the Royal Microscopical Society* (edited by Prof. F. Jeffrey Bell), which gives summaries of recent researches; the *Quarterly Journal of Microscopical Science* (edited by Profs. E. Ray Lankester, Klein, Sedgwick, and Milnes Marshall); and of course *Nature*, in which summaries and discussions are often to be found. More popular journals are the *American Naturalist* and *Natural Science*.

Of all elementary books the best to begin with are two volumes by A. B. Buckley, *Life and her Children* (backboneless animals), and *Winners in Life's Race* (backboned animals); but I shall now mention other ways of beginning.

B. Natural History.

"Certain dreadfully scientific persons, who call themselves by the name of naturalists, seem to consider zoology and comparative anatomy as convertible terms. When they see a creature new to them, they are seized with a burning desire to cut it up, to analyse it, to get it under the microscope, to publish a learned book about it which no one can read without an expensive Greek lexicon, and to put up its remains in cells and bottles. They delight in an abnormal hæmapophysis; they pin their faith on a pterygoid process; they stake their reputation on the number of tubercules on a second molar tooth; and they quarrel with each other about a notch on the basisphenoid bone." Thus, in a breezy way, did the Rev. J. G. Wood laugh at the morphological zoologists. But his good-humoured criticism is apt to be misleading. For if science, as such, be justifiable, the work of the anatomist is warranted as part of it, and is neither less nor more valuable than that of the field naturalist. We may criticise the details of the anatomist's analysis, we may believe that his discipline is often pressed unnaturally upon students, we may beseech him to be less pedantic; but to remind him that the study of structure requires to be supplemented by the study of life is like reminding the field naturalist that animals have bones and muscles. Both are true statements, but somewhat obvious.

The zoologist has deliberately given himself up to analysis, and if the world is to become translucent to us, we must include within our knowledge what he can tell us about the structure and activities of animals, alike as unities and as complex combinations of organs, tissues, and cells. Let us agree to call this serious study, including the morphological and physiological aspects which we have already explained, "zoology." We must acknowledge that few of us can become zoological experts. But let not this hinder us from perceiving that it is not difficult to understand towards what end and by what method Linnæus and Cuvier, Bichât and Claude Bernard, and the other great masters worked; nor let it deter us from using all natural opportunities of practically observing the forms and powers of animal life. We shall soon feel that "zoology" is neither less interesting nor less essential than the work of the field naturalist, we shall recognise that its terminology is not more complex than that of seamanship, and we may even admit that from clear zoological thinking our contemplation of nature acquires an additional intensity of emotion. "Tout naturaliste cachait plus ou moins un amateur d'idylles ou d'éclogues." What Hamerton says with reference to an artist's education applies also to the student of science: "The harm is not in the study (of plants), it is in the forgetfulness of large relations to which this minute observation of nature has occasionally led those who were addicted to it." Zoologists are not the only workers who sometimes lose their sense of perspective.

Now, however, I would address those who have little time or opportunity for "zoology," but who have an interest in the life and habits of animals, and desire to appreciate these more thoroughly. This knowledge of animals as personalities — in struggle and friendliness, in hate and love, in birth and death, I would call "natural history," in contrast to analytic "zoology" on the one hand, and generalising "biology" on the other. For I restrict the latter term to the general theory of life—its nature and origin, its growth and continuance. It matters little what names are given to these three aspects of the study of animal life; thus what I call "Natural History" Prof. Ray Lankester calls more precisely "Bionomics"; but it is important to recognise all the three as essential, and to cease from drawing prejudiced comparisons between them.

Their relations may be summarised as follows :—

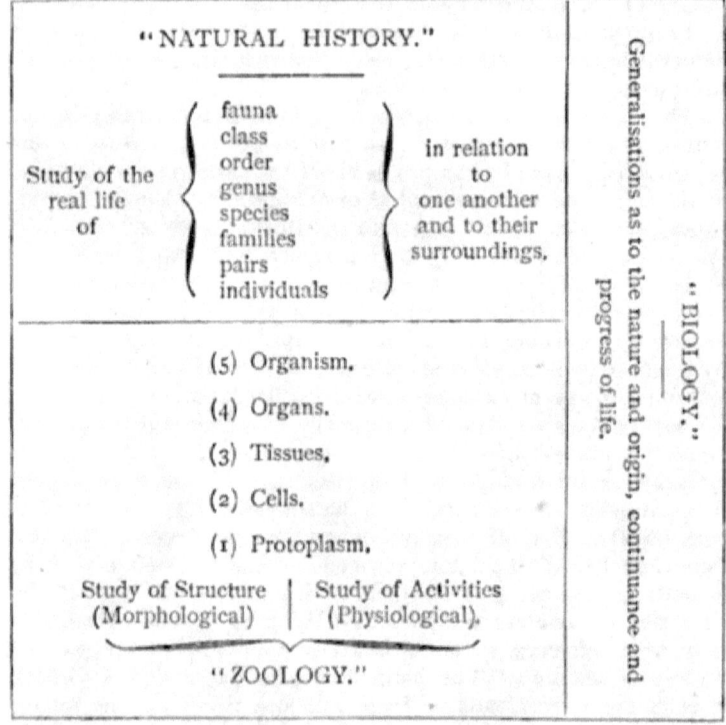

To those interested in "Natural History," there is little need to give the primary word of counsel "OBSERVE," for to do so is their delight; nor do they need to be told that sympathetic feeling with animals, delight in their harmonious beauty, and poetical justice of insight which recognises their personality, are qualities of a true naturalist, as every one will allow, except those who are given up to the idolatry of that fiction called "pure science."

There is a maniacal covetousness of knowledge which one has no pleasure in encouraging. We do *not* want to know all that is contained even in Chambers's *Encyclopædia*, though we wish to gain the power of understanding, realising, and enjoying the various aspects of the world around us. We do *not* wish brains laden with chemistry and physics, astronomy and geology, botany and zoology, and other sciences, though we would have our eyes lightened so that we may see into the heart of things, our brains cleared so that we may understand what is known and unknown when we are brought naturally in face of problems, and our emotions purified so that we may feel more and more fully the joy of life. Therefore I would, in the name of education, urge students to begin naturally, with what interests them, with the near at hand, with the practically important. A circuitous course of study, followed with natural eagerness, will lead to better results than the most logical of programmes if that take no root in the life of the student.

Let me suggest some of these indirect ways of beginning. Begin with domesticated animals and their history. See Darwin's *Variation of Animals and Plants under Domestication* (1868), etc. Concentrate your attention on some common animals. See, for instance, Darwin's *Formation of Vegetable Mould through the action of Worms* (1881); Mivart's *Frog* (Nature Series, London); Huxley's *Crayfish* (Internat. Sci. Series, London); M'Cook's *North American Spiders* (2 vols., Philadelphia, 1889-90); F. Cheshire's *Bees and Bee-keeping* (vol. i., Lond., 1886); Lubbock's *Ants, Bees, and Wasps* (Internat. Sci. Series, London); Flower's *Horse* (Lond., 1891).

Enjoy your seaside holiday. See Charles Kingsley's *Glaucus*; J. G. Wood's *Common Objects of the Sea-Shore* (1857); P. H. Gosse's *Manual of Marine Zoology* (1856), and *Tenby*; G. H. Lewes's *Seaside Studies* (Edin. 1858); L. Fredéricq, *La Lutte pour l'existence chez les Animaux Marins* (Paris, 1889).

Form an aquarium. See J. G. Wood's *Fresh and Salt Water Aquarium*; P. H. Gosse, *The Aquarium* (1854), and many similar works.

Begin a naturalist's year-book. See the *Naturalist's Diary* by Roberts; the *Field Naturalist's Handbook*, by J. G. and Th. Wood (Lond., 1879); and K. Russ, *Das heimische Naturleben im Kreislauf des Jahres; Ein Jahrbuch der Natur*. (Berlin, 1889).

Observe the animals you see on your country walks. See J. G. Wood's *Common Objects of the Country* (1858), *The Brook and its Banks* (1889); *Life of a Scotch Naturalist*, Thomas Edward, by Samuel Smiles; *The Moor and the Loch*, by J. Colquhoun (Edin. 1840, 8th ed. 1878); *Wild Sports and Natural History of the Highlands*, by Charles St. John (Lond., illust. ed., 1878); *Woodland, Moor, and Stream*, edited by J. A. Owen (Lond., 1889); W. Marshall, *Spaziergänge eines Naturforschers* (Leipzig, 1888); Lloyd Morgan's *Sketches of Animal Life* (Lond., 1892), etc. etc.

Another natural way of beginning is to **work** out some subject which attracts you. It becomes a centre round which a crystal grows. Muybridge's photographic demonstrations of animal locomotion have interested us in the flight of birds, let us follow this up by observation and by reading, *e.g.*, Ruskin's *Love's Meinie* (**1881**); Pettigrew's *Animal Locomotion* (Internat. Sci. Series, **1873**); Marey's *Animal Mechanism* (Internat. Sci. Series, 1874); **Marey's** *Le Vol des Oiseaux* (Paris, 1890).

The colours of animals appeal to many people. Read E. B. Poulton's volume (1890) in the Internat. Sci. Series, and **Grant Allen's** *Colour Sense*, and F. E. Beddard's *Animal Colouration* (Lond., 1892).

The relations between plants and animals are entrancingly interesting. Watch the bees and other **insects** in their flight, and read Darwin's **volumes** on the *Fertilisation of Orchids* (1862) and on *Cross-Fertilisation* (1876); Hermann Müller's *Fertilisation of Flowers* (transl. by Prof. D'Arcy Thompson, Lond., 1883); Kerner's *Flowers and their Unbidden Guests*; the articles on "Insectivorous Plants," in *Encyclop. Britannica*, and in Chambers's *Encyclop.*, or Darwin's work (1875).

Again, many of **us** are directly interested in foreign countries. Let the practical interest broaden, it naturally becomes geographical and physiographical, and extends to the natural history of the region. No more pleasant and sane way of learning about the ways and distribution of animals could be suggested than that which follows as a gradual extension of physiographical knowledge. See Dr. H. R. Mill's *Realm of Nature*, and the following samples from the long list of books by exploring naturalists :—

A. Agassiz, *Three Cruises of the "Blake"* (Boston and New York, 1888).
S. W. Baker, *Wild Beasts and Ways: Reminiscences of Europe, Asia, Africa, and America* (London, 1890).
H. W. Bates, *Naturalist on the **Amazons*** (5th ed., London, 1884).
T. Belt, *Naturalist in Nicaragua* (2nd ed., London, 1888).
W. T. Blanford, *Observations on Geology and Zoology of Abyssinia* (Lond., 1870).
P. B. Du Chaillu, *Explorations and Adventures in Equatorial Africa*, (Lond., 1861); *Ashango Land* (1867).
R. O. Cunningham, *Notes on the Natural History of the Straits of Magellan* (Edin., 1871).
Darwin, *Voyage of the "Beagle"* (1844, new ed. 1890).
H. Drummond, *Tropical Africa* (Lond., 1888).
H. O. Forbes, *A Naturalist's Wanderings in the Eastern Archipelago* (Lond., 1885).
Guillemard, *Cruise of the "Marchesa"* (**Lond.**, 1886).

A. Heilprin, *The Bermuda Islands* (Philadelphia, 1889).
S. J. Hickson, *A Naturalist in North Celebes* (London, 1889).
W. H. Hudson, *The Naturalist in La Plata* (Lond., 1892).
A. v. Humboldt, *Travels to the Equinoctial Regions of America* ; *Aspects of Nature* (Trans. 1849) ; *Cosmos* (Trans. 1849-58).
Lumholtz, *Among Cannibals* (Lond., 1889).
H. N. Moseley, *Notes by a Naturalist on the* "*Challenger*" (Lond., 1879, new ed. Lond., 1892).
A. E. Nordenskiöld, *Voyage of the* "*Vega*" (Lond., 1881).
F. Oates, ed. by C. G. Oates, *Matabele Land and the Victoria Falls, a Naturalist's Wanderings in the Interior of S. Africa* (1881).
N. M. Przewalski, *Wissenschaftliche Resultate der nach Centralasien unternommenen Reisen* (Leipzig, 1889).
H. Seebohm, *Siberia in Europe* (Lond., 1880) ; and *Siberia in Asia* (1882).
J. E. Tennent, *Natural History of Ceylon* (Lond., 1861).
Wyville Thomson, *The Depths of the Sea* (Lond., 1873) ; *Narrative of the Voyage of the* "*Challenger*" (1885). Cf. A. de Folin, *Sous les Mers* (Paris, 1887) ; H. Filhol, *La Vie au fond des Mers* (Paris, 1886) ; W. Marshall, *Die Tiefsee und ihr Leben* (Leipzig, 1888).
Tristram, *The Flora and Fauna of Palestine*.
Tschudi, *Thierleben der Alpenwelt*.
A. R. Wallace, *Malay Archipelago* (**Lond.,** 1869) ; *Tropical Nature* (**1878**) ; *Island Life* (1880).
Ch. Waterton, *Wanderings in South America* (ed. by J. G. Wood, 1878).
C. M. Woodford, *Naturalist among the Head-hunters* (London, 1890).

Prominent among those who have helped many to realise the marvel and beauty of nature, a widely-felt gratitude ranks Gilbert White, Henry Thoreau, Charles Kingsley, Richard Jefferies, J. G. Wood, John Ruskin, and John Burroughs.

GILBERT WHITE (1720-1793) is known to all as the author of *The Natural History and Antiquities of Selborne, in the County of Southampton* (1788), a book consisting of a series of letters addressed to a few friends. A good edition is that by J. E. Harting (6th ed., London, 1888), but there is a cheaper one, edited by Richard Jefferies, in the Camelot Series.

HENRY THOREAU (1817-1862), the author of *Walden*, *A Week on Concord*, and other much-loved books.

CHARLES KINGSLEY (1819-1875). See his *Glaucus* (Lond., 1854) ; *Water-Babies* ; and popular lectures.

RICHARD JEFFERIES (1848-1887).

See *The Eulogy of Richard Jefferies*, by Walter Besant (London, 1888), and the following works, some of which are published in cheap editions: *The Gamekeeper at Home* (1878); *Wild Life in a Southern County* (1879); *The Amateur Poacher* (1880); *Round about a Great Estate* (1881); *Nature near London* (1883); *Life of the Fields* (1884); **Red** *Deer* (1884); *The Open Air* (1885).

J. G. WOOD, whom we have lately lost, has done more than any other to popularise natural history in Britain.

See *Life of J. G. Wood*, by his son, Theodore Wood (Lond., 1890); *My Feathered Friends* (1856); *Common Objects of the Seashore* (1857); *Common Objects of the Country* (1858); his large *Natural History* (1859-63); *Glimpses into Petland* (1862); *Homes without Hands* (1864); *The Dominion of Man* (1887); and other works.

JOHN RUSKIN. See the *Eagle's Nest, Queen of the Air, Love's Meinie, Proserpina, Deucalion,* and *Ethics of the Dust.*

JOHN BURROUGHS.

See the neat shilling editions of **Wake** *Robin* (1871), *Winter Sunshine* (1875), *Birds and Poets* (1877), *Locusts and Wild Honey* (1879), *Pepacton* (1881), *Fresh Fields* (1884), *Signs and Seasons* (1886).

See also :—

GRANT ALLEN, *The Evolutionist at Large; Vignettes from Nature*, etc.

FRANK BUCKLAND, *Curiosities of Natural History* (London, 1872-77), and his *Life.*

P. H. GOSSE. *Romance of Natural History* (London, 1860-61).

P. G. HAMERTON, *Chapters on Animals; The Sylvan Year* (3rd ed., London, 1883).

W. KIRBY AND W. SPENCE, *Introduction to Entomology* (London, 1815).

F. A. KNIGHT, *By Leafy Ways; Idylls of the Field* (London, 1889).

PHIL ROBINSON, *The Poet's Birds* (London, 1883); and *The Poet's Beasts* (London, 1885).

ANDREW WILSON, *Leaves from a Naturalist's Note-Books; Chapters on Evolution*, etc.

C. Biology.

Having offered counsel to those who would study the literature of Zoology and of Natural History, I shall complete my task of giving advice by addressing those who are strong enough to inquire into the nature, continuance, and progress of life. It is to students of mature years that this "biological" study is most natural, for young folks should be left to see and enjoy as much as possible, till theories grow in them as naturally as "wisdom teeth." This also should be noted in regard to the study of evolution and the related problems of biology, that though all the generalisations reached must be based on the research and observation of zoologists, botanists, and naturalists, and are seldom fully appreciated by those who have little personal acquaintance with the facts, yet sound and useful conclusions may be, and often are, obtained by those who have had no discipline in concrete scientific work.

Besides the general question of organic evolution there are special subjects which the student of biology must learn to think about: Protoplasm, or "the physical basis of life;" Reproduction, Sex, and Heredity, or "the continuance of life;" and Animal Intelligence, or "the growth of mind." Before passing to the literature on these subjects, it may be noted that there are two general works of pioneering importance, namely, Herbert Spencer's *Principles of Biology* (2 vols., Lond., 1864-66), and Ernst Haeckel's *Generelle Morphologie* (2 vols., Berlin, 1866).

Protoplasm.—Of this the student should learn how little we know. Yet this is not very easy, since the most important recent contributions, such as those of Professors Hering and Gaskell, are inaccessible to most. The gist of the matter, however, may be got hold of by reading: (*a*) three articles in the *Encyclopædia Britannica*, "Physiology" (Prof. M. Foster), "Protoplasm" (Prof. P. Geddes), and "Protozoa"—the large type—(Prof. E. Ray Lankester); (*b*) the Presidential Address to the Biological Section of the British Association, 1889, by Prof. Burdon Sanderson (*Nature*, xl., September 1889, pp. 521-526); and (*c*) the article "Protoplasm" in the new edition of Chambers's *Encyclopædia*. Of the abundant literature on the philosophical questions which the scientific conception of living matter raises, I shall mention Huxley's address on "The Physical Basis of Life," published among his collected essays; Hutchison Stirling's tract, "As regards Protoplasm;" the chapter on "Vitalism" in Bunge's *Physiological Chemistry* (translated, London, 1890).

Reproduction, Sex, and Heredity.—For adult students, and no others should be encouraged to face the responsibility of inquiry into such matters, the most convenient introduction will be found in *The Evolution of Sex* (Contemporary Science Series, Lond., 1889), by Prof. Geddes and myself. In that work there are references to others. A survey of modern opinions and conclusions in regard to heredity may be obtained from the article in Chambers's *Encyclopædia*, whence the student will pass unbiassed to the essays of Weismann, *Papers on Heredity and Kindred Subjects* (translated by E. B. Poulton, S. Schönland, and A. E. Shipley, Oxford, 1889), to the works of Francis Galton, especially his *Natural Inheritance* (Lond., 1889), and to other important books mentioned in the article referred to.

Animal Intelligence.—A recent work by Professor C. Lloyd Morgan, *Animal Life and Intelligence* (Lond., 1890), supplies the best introduction to those interesting questions in the discussion of which the biologist becomes a psychologist. The most reliable treasury of facts is certainly G. J. Romanes's *Animal Intelligence* (Internat. Sci. Series, 4th ed., Lond., 1886), to which may be added Couch's *Illustrations of Instinct* (1847), Lauder Lindsay's *Mind in Animals* (1879), Büchner's *Aus dem Geistesleben der Thiere* (2nd ed., Berlin, 1877) and *Liebe und Liebesleben in der Thierwelt* (Berlin, 1879); Max Perty, *Ueber das Seelenleben der Thiere* (Leipzig, 1876); Houzeau, *Des Facultés mentales des Animaux* (Brussels, 1872). Of unique value is the work of A. Espinas, *Des Sociétés Animales, Étude de Psychologie comparée* (Paris, 1877). See also P. Girod, *Les Sociétés chez les animaux* (Paris, 1890). I should also mention that Brehm's *Thierleben* (1863-69), a great work now in process of re-edition (10 vols., Leipzig), is a marvellous treasury of information in regard to the ways and wisdom of animals, and that we have in Verworn's *Psycho-Physiologische Protisten Studien* (Jena, 1889) a very interesting and important study of the dawn of an inner life in the simplest animals or Protozoa. Of the ingenious work of animals, an admirably terse description is given in F. Houssay's *Les Industries des Animaux* (Paris, 1889). For theories of instinct, see especially Romanes, *Mental Evolution in Animals* (Lond., 1883); Darwin, *Origin of Species*; Wallace, *Contributions to the Theory of Natural Selection*; Spencer, *Principles of Psychology* and *Principles of Biology*; G. H. Lewes, *Problems of Life and Mind* (Lond., 1874-79); Samuel Butler, *Life and Habit* (Lond., 1878); J. J. Murphy, *Habit and Intelligence*; E. von Hartmann, *Das Unbewusste vom Standpunkte der Physiologie und Descendenztheorie* (2nd ed., Berlin, 1877); Schneider, *Der Thierische Wille* (Leipzig, 1880); Eimer, *Organic Evolution*; Weismann, *Papers on Heredity*.

The Fact of Organic Evolution.—The student's first task in regard to Evolution is to make himself acquainted with the arguments which show that the animals and plants now alive are descended from simpler ancestors, these from still simpler, and so on back into the mists of life's beginnings. To realise that the present is child of the past is to realise the fact of Evolution, and the surest way to grasp the biological verification of this fact is to undertake a course of practical study. Failing this, we must, I suppose, read up the subject. Romanes's *Evidences of Evolution* (Nature Series, Lond.) gives a convenient statement of the case, and his Rosebery Lectures will be more exhaustive. Clodd's *Story of Creation: a plain account of Evolution* (Lond., 1888) sums up the evidence in small compass; another very terse statement will be found in H. De Varigny's *Experimental Evolution* (Lond., 1892); Haeckel's *Natural History of Creation* (Berlin, 1868)—the most popular of his works, now in its eighth edition (Jena, 1890)—is available in translation (Lond., 1879); Huxley's *American Addresses* (Lond., 1877) have even greater charm of style; Carus Sterne's *Werden und Vergehen* (3rd ed., Berlin, 1886) is perhaps the best of all popular expositions; while the thorough student will find most satisfaction in the relevant portions of Darwin's *Origin of Species*, and Spencer's *Principles of Biology*.

History of Evolution Theories.—As the idea of Evolution is very ancient, and as it was expounded in relation to animal life by many competent naturalists before Darwin's intellectual coin became current throughout the world, it is unwise that students should restrict their reading to Darwinian and post-Darwinian literature. The student of Evolution should know how Buffon, Erasmus Darwin, Lamarck, Treviranus, the St. Hilaires, Goethe, even Robert Chambers, and many other pre-Darwinians dealt with the problem. Those who desire to preserve their sense of historical justice should read one or more of the following: Huxley's article on "Evolution" in the *Encyclopædia Britannica*; Samuel Butler's interesting volume on *Evolution Old and New* (Lond., 1879); Perrier's *Philosophie Zoologique avant Darwin* (Paris, 1884); the historical chapters of Haeckel's *Natural History of Creation*; Carus's *Geschichte der Zoologie*, and some other historical works already referred to (p. 355); A. de Candolle's *Histoire des Sciences et des Savants dépuis deux Siècles* (Genève, Bâle, 1883); Carus Sterne's (Ernst Krause's) excellent work, *Die Allgemeine Weltanschauung* (Stuttgart, 1889); De Quatrefages, *Charles Darwin et ses précurseurs français* (Paris, 1870).

Darwinism.—The best account of the Darwinian theory of Evolution, especially of the theory of natural selection which Charles Darwin and Alfred Russel Wallace independently elaborated, is Wallace's *Darwinism* (Lond., 1889). From this the student will naturally pass to the works of Darwin himself—*The Origin of Species by means of Natural Selection ; or, the Preservation of Favoured Races in the Struggle for Life* (Lond., 1859); *The Variation of Animals and Plants under Domestication* (2 vols., Lond., 1868); *The Descent of Man, and Selection in Relation to Sex* (Lond., 1871), etc. ; the earlier works of Wallace, especially his *Contributions to the Theory of Natural Selection* (Lond., 1871); Spencer's **Principles of** *Biology—cf.* his articles on " The Factors of Organic **Evolution** " (*Nineteenth Century*, 1886); Haeckel's *Generelle Morphologie*, and *Natural* **History of** *Creation*. As a popular account of Darwin's life and work, Grant Allen's *Charles Darwin* (English Worthies Series, 3rd ed., Lond., 1886) has a deserved popularity; G. T. Bettany's similar work **(Great** Writers Series, Lond., 1886) has a very valuable biblio**graphy;** but for full personal and historical details reference must be made to the *Life and Letters of Charles Darwin*, by his son Francis Darwin (3 vols., Lond., 1887).

Recent Contributions to the Theory of Evolution.— At the present time there is much discussion in regard to the **factors of** organic Evolution. The theory of Evolution is still being evolved; there is a struggle between opinions. On the one hand, many naturalists are more Darwinian than Darwin was, —that **is** to say, they lay more exclusive emphasis upon the theory of natural selection; on the other hand, not a few are less Darwinian than Darwin was, and emphasise factors of Evolution and aspects of Evolution which Darwin regarded as of minor importance.

Of those who are more Darwinian than Darwin, I may cite as representative : Alfred Russel Wallace who, in his *Darwinism*, subjects Darwin's subsidiary theory of sexual selection to destructive criticism; August Weismann who, in his Essays on Heredity, denies the transmissibility of characters acquired by the individual organism, **as** the results of use **or** disuse or of external influence ; and E. Ray Lankester, see his **article** " Zoology " in the *Encyclopædia Britannica*, and his **work** on the *Advancement of Science* (Lond., 1890). The student should also read an article by Prof. Huxley, "The Struggle for Existence, and its Bearing upon Man " in the *Nineteenth Century*, Feb. 1888.

See also :—

Samuel Butler, *Evolution Old and New* (Lond., 1879), *Luck or Cunning* (Lond., 1887), and other works.

Prof. E. D. Cope, *Origin of the Fittest* (New York, 1887).

Prof. G. H. T. Eimer, *Organic Evolution, as the Result of the Inheritance of Acquired Characters, according to the Laws of Organic Growth* (Jena, 1888). Trans. by J. T. Cunningham (Lond., 1890).

Some of the "Best Books" on Animal Life

Prof. T. Fiske, *Outlines of Cosmic Philosophy* (Lond., 1874), *Darwinism, and other Essays* (Lond., 1875).
Prof. P. Geddes, Article "Variation and Selection," *Encyclopædia Britannica*; "Evolution," *Chambers's Encyclopædia*, new ed. Cf. *The Evolution of Sex*, and forthcoming work on *Evolution, Organic and Social*.
E. Gilou, *La Lutte pour le Bien-être* (1890).
Rev. J. T. Gulick, *Divergent Evolution, through Cumulative Segregation* (Journ. Linn. Soc. xx., 1888).
P. Kropotkine, "Mutual Aid among Animals," *Nineteenth Century* (Sept. and Nov. 1890).
Lanessan, *La Lutte pour l' Existence et l' Association pour la Lutte* (Paris, 1882).
Prof. St. George Mivart, *The Genesis of Species* (Lond., 1871), *Lessons from Nature* (Lond., 1876), *On Truth* (Lond., 1889).
Prof. C. Lloyd Morgan, *Animal Life and Intelligence* (Lond., 1890).
Prof. C. V. Nägeli, *Mechanisch - physiologische Abstammungslehre* (München and Leipzig, 1884).
Prof. A. S. Packard, Introduction to the *Standard* or *Riverside Natural History* (New York and Lond., 1885).
Dr. G. J. Romanes, *Physiological Selection* (Journ. Linn. Soc. xix., 1886), and forthcoming Rosebery Lectures on the Philosophy of Natural History.
Prof. K. Semper, *The Natural Conditions of Existence as they affect Animal Life* (Internat. Sci. Series, Lond., 1881).
Dr. J. B. Sutton, *An Introduction to General Pathology* (Lond., 1886). *Evolution and Disease* (Contempor. Sci. Series, Lond., 1890).

INDEX

ABSORPTION, 145
Acacias guarded by ants, 29, 30
Acquired characters, 329-336
Actions, automatic, 155
  habitual, 155
  innate, 155
  intelligent, 155
Alternation of generations, 189
Amœba, 213
Amphibians, 9, 256, 257
  parental care among, **110, 111**
Amphioxus, 252
Angler-fish, 118
Animalculists, 191
Animals, everyday life of, 1-124
  domestic life of, 95, 116
  industries of, 117-124
  life-history of, **184-203**
  past history of, **204-209**
  social life of, 67-94
  and plants, resemblances and contrasts, 167-171
  relation of simplest to more complex, 171-174
Annelids, 231-234
Antlers, 279
Ants, 78-84
  and aphides, **119, 120**
  and plants, 29
Aphides, **82,** 312
  multiplication of, 38
Arachnida, 243
Archoplasm, 183

Aristotle, 283, 284
Armour of animals, 34, 35
Artemia, 310, 311
Arthropods, 10, **238**
Atavism, 322
Autotomy, 64-66
Axolotl, 309

BACKBONED animals, 9, 222-247
Backboneless animals, 9, 10, 248-272
Bacteria, **21, 22**
Balance of nature, 19-21
Balanoglossus, 9, 249, 250
Bathybius, 219
Beauty of animals, 15-17
Beavers, **25,** 74, 75
Bees, 78-84 *75-78*
Biology, justification of, 34-50
Birds, 9, 264-267
  parental care among, 114, 115
Blind animals, 305
Body, functions of, 144-149
  parts of, 174-183
Books, 351-369
Boring animals, 25
Bower birds, 98
Brachiopoda, **235**
Brine-shrimp, **310, 311**
Buffon, **286**

CADDIS worms, **61**
Carbohydrates, **134**

Caterpillars, 50, 51
Cats and clover, 29
Cave-animals, 334
Cell-division, 158-183
Cells, 128, 147, 179-183
Centipedes, 241
Cestoda, 229
Chætopoda, 231-233
Challenger Expedition, 5, 6
Chamæleons, 52
Chemical elements, 135
    influences in environment, 309, 313
Circulation, 146
Classification of animals, 8-11
Cœlenterates, 222-228
Cold, effect of, 313
Colonies, 70, 71
Colour-change, 52, **53**
Colouring, protective, 48, **49**
    variable, 49-51
Colours of animals, **49-53**
    of flat-fishes, 315
Commensalism, 68, 69
Competition, internal, 67
Concealment of animals, 47
Conjugation, 214
Consciousness, 150-**152**
Co-operation, 69
Corals, 26, 27, 227
Coral snakes, 59
Courtship of birds, 96
    mammals, 96
    spiders, 101-105
Crabs, masking of, 61, 62
    and sea-anemones, 68, 69
Cranes, gregarious life of, 73
Crayfish, 25
Crocodilians, 263, **264**
Cruelty of nature, 43-45
Crustacea, 239, 240
    life-history of, 198-202
Cuckoo, 114, 115
Cuttlefish, 52, 66
Cyclostomata, 252

DARWIN, Charles, 292-296
    Erasmus, 288, 289

Deep-sea fishes, 256
    life, 6
Descent of man, 341-345
Desiccation, 41-43
Digestion, 145
Distribution of animals, 3-8
Disuse, results of, 305, 306
Division of labour, 69-71, 143, 144
Dormant **life**, 41-43
Drought, **effect of**, 41-43

EARTHWORMS, 22-24
Echinoderms, 10, 65, 66, 235-238
Ectoderm, 196
Eggs, 191, 192
Elaps, 59
Elephant **hawk-moth**, **59**
Encystation, 41
Endoderm, 196
Environment, **306-319**
Ephemerides, 106, 107
Epiblast, 196
Epigenesis, **324**
Evolution, evidences of, 273-281
    factors of, 299-302
    theories, history of, 282-301
    of sex, 188
Extinct types, **206, 207**

FAMILY, evolution **of**, **91**
    life, 91
Fats, 134
Feigning death, 66
Fertilisation, **193-195**
Filial regression, 338
Fishes, 9, 253-256
    parental care among, 109, 110
Flight of birds, 123, 124
Flowers and insects, 28, 29
Flukes, 229
Food, influence of, 310-313
Freshwater fauna, 6-8
Friar-birds, 59
Frog, 258
Function, influence of, 303

GASTRÆA theory, 197

Index

Gastrula, 195, 196
Genealogical tree, 12, 13
Geological record, imperfection of, 205
Germ-plasma, 328
Giant reptiles, 259
Glow-worm, courtship of, 100
Gregarines, 211
Gregarious animals, 71-74
Grouse attacked by weasel, 40

HABITAT, change of, 47
Habitual actions, 155
Haeckel, 298
Hagfish, 253
Halcyon, 116
Hatteria, 260
Heat, influence of, 313
Heredity, 320-339
Hermaphroditism, 188
Hermit-crabs, masking of, 63
Hirudinea, 234
Homes, making of, 121-123
Hornbill, brooding of, 114
Horse, pedigree of, 278
Hunting, 118, 119
Huxley, 298
Hydractinia, 69, 70
Hypoblast, 196

ICHNEUMON flies, 64
Idealism, 142
Impressions, 151
Industries of animals, 116-124
Infusorians, 211
 multiplication of, 38
Innate actions, 155
Insects, 241-243
 parental care of, 108
 and flowers, 28
Instinct, 153-166
 origin of, 163-166
Instincts defined, 11
 incomplete, 158
 mixed, 163
 primary, 163
 secondary, 163
Insulation of animals, 46

Intelligence, lapse of, 166
Intelligent actions, 155
Iron, importance of, 19
Isolation, 300, 301
Ivory, 31

JELLYFISH, 226

KALLIMA, 53
Kidneys, work of, 145

LAMARCK, 289-292
Lamprey, 252
Lancelet, 9, 252
Land animals, 8
Leaf insects, 54
Leeches, 234
Lemming, Ross's, 50
Lemurs, 46
Life, chemical elements of, 135-137
 energy of, 127
 haunts of, 3-8
 machinery of, 130, 131
 origin of, 140-142, 280
 struggle of, 32-45
 variety of, 3
 wealth of, 1-17
Light, influence of, 315, 316
Liver, work of, 145
Living matter, 131-135
Lizards, 260
Love of mates, 90, 91, 96
 and care for offspring, 105-116
 and death, 106
Luciola, courtship of, 100
Lucretius, 284, 285

MACROPOD, parental care of, 110
Mammals, 9, 267-271
Man as a social person, 94
 considered zoologically, 340-346
Marine life, 3-6
Marsupials, 46
Masking, 61-63
Materialism, 141, 142

Mates, love of, 90, 91, 96
Mayflies, 106, 107
Metamorphosis of Insects, 243
Mesoderm or mesoblast, 196
Migration of birds, 74
Millepedes, 241
Mimicry, 57-61
Mites, desiccation of, 41, 42
Molluscs, 10, 243-247
Monkeys, 270, 271, 341
   gregarious life of, 71
Monogamous mammals, 96
Moss-insect, 55
Moulting, 315
Movement, 144
Movements of animals, 123, 124
Mud-fish, 8
Mygale, 36
Myriapoda, 241

NATURAL selection, 295
Nematoda, 231
Nemerteans, 230
Nervous system, 148
Nudibranchs, 56
Number of animals, 14, 15
Nutrition, 144
Nutritive relations, 27, 28

ODOURS and sexual attraction, 105
Offspring, care for, 105-116
Ontogeny, 203
Ooze, 220
Organic continuity, 203, 326-329
Organs, 175
   change of function of, 178
   classification of, 178
   correlation of, 176
   order of appearance of, 175, 176
   rudimentary, 178, 179
   substitution of, 178
Orioles, 59
Ovists, 191
Ovum, 191-193
   theory, 196
Oysters, mortality of, 43

PALÆONTOLOGICAL series, 206
Palæontology, 204-209
Pangenesis, 324
Parasitic worms, 229-231
Parasitism, 47, 48
Parthenogenesis, 189-193
Partnerships among animals, 68, 69
Perception, 151
Peripatus, 10, 240
Phasmidæ, 53
Phenacodus, 269, 270
Phyllopteryx, 54
Phylogeny, 203
Physiology, 125-152
Pigeon, 275, 276
Pineal body, 260
Plants and animals, 19, 20, 28-31, 168-171
Polar globules, 193
Polyzoa, 235
Preformation theories, 191, 324
Pressures, effect of, 300
Protective resemblance, 53
Proteids, 134, 135
Protomyxa, 212
Protoplasm, 131-135
Protopterus, 41
Protozoa, 11, 210-221
   colonial, 173, 174
   classes of, 211
   life of, 214
   "immortality" of, 172
   psychical life of, 215-218
   structure of, 213
   and Metazoa, transition between, 88, 89, 171-174
Psychology, 149
Pupæ of caterpillars, 50
Puss-moth, 63, 64

RADIANT energy, influence of, 313-316
Recapitulation, 197, 279
Reflex actions, 155
Reproduction, 184-190
Reptiles, 9, 259-264
Respiration, 146

Index

Reversion, **322**
Rhizopods, **212**
Rotifers, **7**, 42, 234
Round-mouths, 9, 252
Rudimentary organs, 277

SACCOPHORA, 62
Sacculina, 48
Sea-horse, parental care of, **110**
Seasonal dimorphism, 314
Segmentation, 195
Sensations, 151
Sex, 96
Sexual reproduction, 186-188
  selection, 98
Shells of molluscs, **243**
Shepherding, 119, **120**
Shifts for a living, **46-66**
Skunk, 55
Snails and plants, **30**
Snakes, 260-263
Social inheritance, 337-339
  life of animals, 67-94
  organism, 93, **94**
Societies, evolution of, **87**
Song of birds, 96
Spencer, 297, 298
Spermatozoon, 192, **193**
Sphex, 121
Spiders, courtship of, 101-105
  bird-catching, 36
Sponges, 11, 222
Spongilla, 186
Spring, biology of, 95, 96
Starfish, 235

Stickleback, courtship of, 99
  parental care of, 109, **110**, **122**
Stinging-animals, 11, **223-228**
Storing, 120, 121
Struggle for existence, 32-45
Surrender of parts, 64-66
Symbiosis, 69

TAPEWORMS, 229
Termites, 24, 84-87
Tissues, 179, 180
Tortoises, 263
Toxotes, 118
Trematoda, 229
Tunicates, 9, 250, 251
Turbellaria, 228

VARIATION, 299
Vertebrata, characters of, 19, 248, 249
Vital force, 19
Vivarium, 20, 21
Volvox, 187

WALLACE, 296, 297
Warning colours, 55, 56
Weapons of animals, 34
Web of Life, 18-31
White Ants. *See* Termites
Worms, 10, **11**, **228-235**

YOLK, 195

ZOOLOGY, history of, 352-357

THE END

Printed by R. & R. **CLARK**, *Edinburgh*

University Extension Manuals,

Edited by Professor Knight.

The Realm of Nature, a Manual of Physiography. *With Coloured Maps and Illustrations.*—Hugh Robert Mill, University Lecturer, Edinburgh; Librarian to the Royal Geographical Society. 5s.

The Elements of Ethics.—John H. Muirhead, Balliol College, Oxford; Lecturer on Moral Science, Royal Holloway College. 3s.

The Daily Life of the Greeks and Romans.—W. Anderson, Oriel College, Oxford; and Professor of Classics, Firth College, Sheffield.

The History of Astronomy.—Arthur Berry, Fellow of King's College, Cambridge, and of University College, London.

Shakespeare, and his Predecessors in the English Drama.—F. S. Boas, Balliol College, Oxford.

The English Poets, from Blake to Tennyson.—Stopford A. Brooke, Trinity College, Dublin.

The Fine Arts. *With Illustrations.*—Baldwin Brown, Professor of Fine Arts, University of Edinburgh. 3s. 6d.

English Colonization and Empire. *With Coloured Maps and Illustrations.*—A. Caldecott, Fellow of St. John's College, Cambridge. 3s. 6d.

An Introduction to Physical Science.—John Cox, late Warden of Cavendish College; Fellow of Trinity College, Cambridge; Professor of Natural Philosophy, McGill College, Montreal.

The Use and Abuse of Money.—W. Cunningham, D.D., Fellow of Trinity College, Cambridge. 3s.

Outlines of Modern Botany.—Patrick Geddes, Professor of Botany, University College, Dundee. [*In the Press.*

The Jacobean Poets.—Edmund Gosse, Trinity College, Cambridge.

The Literature of France.—H. G. Keene, Universities of Oxford and Calcutta. 3s.

The Philosophy of the Beautiful.—William Knight, Professor of Moral Philosophy and Political Economy, University of St. Andrews. 3s. 6d.

British Dominion in India.—Sir Alfred Lyall, K.C.B., K.C.S.I.

The French Revolution.—C. E. Mallet, Balliol College, Oxford.
[*In the Press.*

The Physiology of the Senses.—John McKendrick, Professor of Physiology, University of Glasgow, and Dr. Snodgrass, Physiological Laboratory, Glasgow. *With Illustrations.* [*In the Press.*

Comparative Religion.—Allan Menzies, Professor of Theology and Biblical Criticism, University of St. Andrews.

Logic, Inductive and Deductive.—William Minto, Professor of Logic and Literature, University of Aberdeen.

The English Novel, from its Origin to Sir W. Scott.—W. A. Raleigh, Professor of English Literature, University College, Liverpool.

Studies in Modern Geology.—R. D. Roberts, Fellow of Clare College, Cambridge; Sec. to the Cambridge and London Extension Syndicates. *With Coloured Maps and Illustrations.* [*In the Press.*

Problems of Political Economy.—M. E. Sadler, Senior Student of Christ Church, Oxford; Secretary to the Oxford University Extension Delegacy.

Psychology, a Historical Sketch.—Andrew Seth, Professor of Logic, University of Edinburgh.

Outlines of English Literature.—Wm. Renton. *With Diagrams.*
[*In the Press.*

JOHN MURRAY, ALBEMARLE STREET.

www.ingramcontent.com/pod-product-compliance
Lightning Source LLC
Chambersburg PA
CBHW030341230426
43664CB00007BA/490